**Penny van Oosterzee** is an Adjunct Professor at James Cook University and has won two Eureka Science Awards and several Whitley Awards for her writing. She has run several private enterprise environmental businesses in tourism, ecosystem services and environmental consulting, and has been a Governor of WWF Australia and a board member of the Federal Biodiversity Advisory Council. She is author of *The Discovery of the Hobbit*, *Dragon Bones* and *A Natural History and Field Guide to Australia's Top End*.

———

'This enchanting, addictive book is an exploration of the all-consuming love affair between its brilliantly articulate, intuitive author and the majestic rainforests of Australia's Wet Tropics that have shaped and focused her life, as they have done for millions of people over thousands of years. It is a perfect blend of prose, prehistory, history and science. I challenge anyone who starts this book to put it down without first devouring it from cover to cover—I couldn't, and I'm still besotted with the storm of thoughts it's created in me.'

**Professor Mike Archer AM, University of New South Wales**

'Driven by a love of the wet tropical forests, Penny van Oosterzee tells their story from deep time to the present, conveying with immediacy the colonial interactions between Indigenous peoples and invading Europeans, and the urgent need for global action to save the forests and humanity.'

**Libby Connors, author of *Warrior***

'*Cloud Land* is a delight to read. I was fascinated by Penny's account of the evolution of forests, their destruction at the hands of European settlers, the role of Indigenous people in managing landscapes and the systematic theft of their land and loss of cultural knowledge.'

**Mike Berwick, former mayor of Douglas Shire**

## Other books by Penny van Oosterzee

*A Natural History and Field Guide to Australia's Top End*

*The Discovery of the Hobbit: The scientific breakthrough that changed the face of human history*

*Dragon Bones: The story of Peking Man*

*Where Worlds Collide: The Wallace Line*

*A Field Guide to Central Australia: A definitive guide to the natural and cultural values of central Australia*

*The Centre*

*Exploring Nature in the Deserts*

# CLOUD LAND

The dramatic story of
Australia's extraordinary
rainforest people
and country

## PENNY VAN OOSTERZEE

ALLEN&UNWIN
SYDNEY·MELBOURNE·AUCKLAND·LONDON

First published in 2023

Allen & Unwin
Cammeraygal Country
83 Alexander Street
Crows Nest NSW 2065
Australia
Phone:  (61 2) 8425 0100
Email:   info@allenandunwin.com
Web:    www.allenandunwin.com

*Allen & Unwin acknowledges the Traditional Owners of the Country on which we
live and work. We pay our respects to all Aboriginal and Torres Strait Islander
Elders, past and present.*

 A catalogue record for this
book is available from the
National Library of Australia

ISBN 978 1 76106 840 9

Author photograph by Martin Willis
Illustrations by William Cooper
Map by Guy Holt
Index by Garry Cousins
Set in 12/16 pt Adobe Garamond Pro by Midland Typesetters, Australia
Printed and bound in Australia by the Opus Group

10 9 8 7 6 5 4 3 2 1

*To Dan, Milla, Luke and Noel,*
*who make it worthwhile*

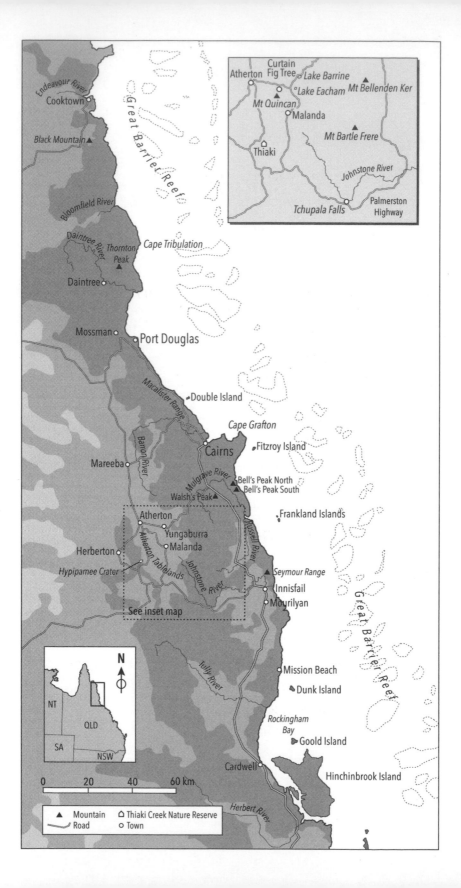

Endeavour River
Cooktown

Black Mountain ▲

Great Barrier Reef

Bloomfield River

Daintree River
Thornton Peak ▲
Cape Tribulation
Daintree

Mossman
Port Douglas

Macalister Range

Double Island

Cape Grafton
Cairns
Fitzroy Island

Barron River

Mareeba

Mulgrave River
Bell's Peak North ▲
Bell's Peak South ▲
Walsh's Peak ▲

Frankland Islands

Atherton
Yungaburra
Malanda

Herberton
Atherton Tablelands
Hypipamee Crater
Johnstone River

Russell River

Seymour Range ▲
Innisfail
Mourilyan

See inset map

Tully River

Mission Beach
Dunk Island

Rockingham Bay
Goold Island

Cardwell
Hinchinbrook Island

Great Barrier Reef

Herbert River

Inset map:

Atherton
Curtain Fig Tree
Lake Barrine
Lake Eacham
Mt Bellenden Ker ▲
Mt Quincan ▲
Malanda
Mt Bartle Frere ▲
Thiaki
Johnstone River
Tchupala Falls
Palmerston Highway

N

NT
QLD
SA
NSW

0    20    40    60 km

▲ Mountain        △ Thiaki Creek Nature Reserve
  Road            ○ Town

# CONTENTS

# PART ONE

## Land of the
## Tree-kangaroo

### Thiaki rainforest

Water tickling, trickling down drip tips, stems, hollows
   and cracks,
branch-trunk armpits tufted with epiphytes,
scratches of tree-kangaroos and possums
damp crevassed and pitted trunk,
mottled with clinging passengers,
ropes of lianas, lichens, beards of moss, filmy ferns
weirs in the rivers of vertical water
flowing to the ocean of life that is soil and soul,
roots like reefs.

Ahh, sun
warmth on a cool canopy,
burning cloud-drenched leaves,
piercing holes in the dragging mist,
etching the forest silhouette
in slanting shadowed rays
clinging beads of water dissipate
from multi-dimensional layers
in a breath.

Penny van Oosterzee

# CHAPTER ONE

# Ambush

My relationship with Thiaki started with an ambush.

It wasn't your usual ambush, but it was an ambush, never-theless. We'd crept along a narrow track that ran atop a volcanic ridge, the sound of crunching gravel under the car's wheels muffled by a narrow strip of trees edging the track. We came to the back of a brick house, which had a neat lawn adorned with a cement girl holding a concrete birdbath. Sheds were stacked with rainforest timber, and cattle yards were perched further ahead at the end of the ridge.

Ridges with level ground were in short supply on the random boneless volcanic landscape of the Atherton Tablelands. A scruffy dingo–dog cross and a couple of cats lounged near the roller-door entrance, lazily observing the proceedings as their owners, Barry and Kirsten Pember, shook our hands in formal welcome and ushered us into a large living space: concrete floor, laminex-surfaced table and metal chairs, no fuss.

Suddenly, we saw something captivating through the large east-facing window. Concealed, up until now, was a jungled landscape with the highest mountain in Queensland, Mount Bartle Frere, sheathed in shifting cloud on the horizon. Draped in mist, the rainforest exhaled a long, slow, yogic-like breath; it created a ghostly play of shapes and forms, like an exhumation of prehistoric creatures from a deep memory. They drifted upwards in sinuous lines, dissipating as they merged into the warming air. The rainforest, etched by the rising sun, was a preternatural presence: aloof, almost sentient in its self-awareness. But with as many individual organisms as cells in the human brain, it was hardly an 'it'. Nor a he or she. I was looking at 'them'.

I looked sideways at the profiles of my husband and son who were slack-jawed in wonder. I guess we'd have to buy them, then: the forest was for sale. I'd discovered them on the internet just a few weeks before. I was aware of the absurdity of it, like a short-lived bacterium in my gut deciding it needed to buy me. Nevertheless, this forest could be ours for the short microbial life left to us.

———

At the bottom of the steep little valley that cradles a gurgling spring-fed stream, we have a water pump squatting in a small rocky hole and covered by the rotund body of an old wringer washing machine to protect it from silt. Steep enough to make walking difficult, Pump Valley was one of the few valleys of Thiaki rainforest spared from logging, so massive trees over 30 metres in height still dapple leaf-littered slopes in hazy light. My husband, Noel, and I were making our way down the 40-degree slope to check the pump, scrambling from one tree trunk to the next to break our fall. We stopped to catch our breath just above the sheer-sided bank of the creek, admiring the green light of deep forest before descending.

Suddenly, there was a loud and sickening snap directly above us, accompanied by a whirr and shower of leaves. We threw our arms above our heads, expecting to be crushed by a large branch complete with a tonne of epiphytes, but instead a chunky tree-kangaroo plummeted to the ground at our feet, so close that it flicked the rim of Noel's hat. I watched in astonishment as the animal landed first on a backbone seemingly reinforced for such an activity, and then rolled the few metres down to the creek bottom where she frenetically found her hind feet and took off, apparently unscathed, up the opposite slope in a frenzy of low-crouching leaps typical of the species. Recently evolved from their rock-wallaby forebears to climb trees, Lumholtz's tree-kangaroos drop to their ancestral ground when they feel threatened.

Still, this is reckless behaviour for a creature weighing up to 10 kilograms. For an instant, I imagined the disbelieving face of a police officer as I explained that a plummeting tree-kangaroo had killed my life-insured husband. I've always thought that Thiaki had a certain attitude that was embodied in stinging trees, lawyer vines, scrub itch, ticks and missiles of dead or rotting branches, not to mention the odd antipodean drop bear.

Thiaki is a 180 hectare scrap of cloud forest, the northern section of a larger 1000-hectare patch, in the Wet Tropics of Australia. The patch is the largest privately owned remnant of rainforest that once cloaked the entire Atherton Tablelands of Far North Queensland. Frequently swathed in mist, Thiaki is a known hotspot for tree-climbing kangaroos, lemuroid ringtail possums, Herbert River ringtail possums and the green ringtail possum. All 13 species of birds that are endemic to northern Queensland rainforests—including cassowaries and golden bowerbirds—grace the place.

Thiaki forest harvests the first runnels of rainwater that eventually collect in two streams; these slip down the irregular valleys and chisel the sides of the dormant three-million-year-old Malanda volcano before merging into Thiaki Creek and then the North Johnstone River that collects water from the southern and south-eastern Atherton Tablelands before rushing out to the Great Barrier Reef at Innisfail. A wistful Greek must have named

Thiaki Creek, and the adjacent Ithaca Creek, in memory of Homer's *The Odyssey*.[1] This ancient Greek epic follows heroic Odysseus, King of Ithaca, who strives against the whim of meddlesome gods to return home after the battle of Troy, where his signature Trojan horse trick wins the war.

But a place less like thirteenth-century BC Ithaca—also called 'Thiaki' by the Ionian locals—with its bare goat-grazed slopes and groves of olives and grapes, you couldn't imagine. The identity of the forlorn Greek is lost to history, as short as European history is in this part of the world where Dreaming and evolution rule.

Being the second work of Western literature, Homer's epic just tips into history. But in the world of Homer, fantasy and reality blend in the bizarre and grotesque. The world is ruled by the supernatural. There are gods and spirits everywhere, creating havoc, having sex with humans, wrecking ships at sea, changing people into pigs, and luring sailors to their death with siren song. You can visit hell and speak to the mournful dead. There are nymphs of springs and rivers, caves and lakes; there are monsters and witches; there are cannibals dashing men's brains on the ground and tearing them limb from limb 'leaving nothing, neither entrails nor flesh, marrow nor bones'.[2]

The homesick Greek must have had a fantastic imagination. Or was he onto something?

———

Once upon a time, tells the Australian Aboriginal cosmology, the land was featureless and bare of life. There were no plants to clothe its nakedness; there was no sound to break its utter silence. For an eternity, this Dreaming world lay flat and unbroken. Ancestral life, hidden under the surface, began to lurch and stumble onto the forbidding landscape to mould, gouge, compress and stretch it with geological force. Ages passed, and creatures of wonderful diversity and phenomenal size emerged and moved across the land.

Everywhere the god-like incarnations roamed, everything they did, and any heroic interaction they had—fighting, fucking, gathering food, resting, eating, participating in ceremonies, giving birth, shitting, dying—spilled life essence on the land and marked it with a natural feature. Shimmering between human and non-human form, these great spirit shapeshifters patterned the land in an elaborate weaving of mythical Dreaming tracks.[3]

Australia's hundreds of Aboriginal countries are uniquely bound by a fine mesh of Dreaming tracks: songlines that stitch and weave the world together into a complex and connected pattern. Anthropologists Catherine and Ronald Berndt[4] suggested that a 'who's who' list of mythical characters from just one Country would today fill a volume; a map of Australia would be scrimshawed with a complicated crisscrossing of mythical tracks and traces, and pierced with galaxies of sites: rock paintings, rocks, lakes, pools, trees, hills and mountains, gorges and waterfalls. All sentient, all made by great ancestors living, dying, fighting and loving, connecting one place to another, one species to another: 'singing up everything';[5] performing rituals, distributing plants, making landforms and water; leaving totems and totemic centres that weld specific human groups with non-humans or with phenomena such as wind and rain. Making Laws for living.

The ancestors have human characteristics: some are kind; some are very bad indeed and make Homer's world look like a kindergarten; some are sacred Lawgivers with hypernatural powers that ignore time and space, and resonate to this day in strong emotions; some are simply ancestors going about their lives. Many travelled from across the sea; some emerged in a particular region and dissolved back into the ground in the same area. Others journeyed across vast distances, leaving a trail of sacred sites across the continent. Some fled to the sky as stars and the moon.

In highly populated countries, where resources were particularly rich, such as those encompassing the modern-day cities of Darwin, Brisbane, Sydney, Melbourne and Perth, the scrimshaw of Dreaming tracks might be comparable to a high-density, imbricated rail network, irregularly gridding the Country and connecting each named stopping point and intersection.[6]

Deborah Bird Rose[7] provided some of the most evocative prose I know to describe for non-Indigenous people how Aboriginal people view Country. Country is a living entity with a consciousness, and a will towards life. People talk about Country in the same way that they would talk about a person: they speak to Country, sing to Country, visit Country, worry about Country, feel sorry for Country and long for Country. People weep for Country. There is no such thing as 'spending a day in the country', no such thing as going out 'into nature'.

I spent a decade living in Central Australia, trying to come to terms with the different worldviews. A Traditional Owner cried with pain and guilt as he explained the theft decades earlier of sacred objects from a cave. It was he who held the ultimate responsibility for the objects; he alone who could use them to make the rivers run with honey, and the Country strong. He blamed himself for the loss—not the thief, not the missionaries, not the murderous cattle barons—and for the deteriorating state of his Country.

Country knows, hears, smells, takes notice, takes care, and is sorry or happy. Country is raw and has a lively sense of humour. I laughed with the women, their gestures raunchy and their ample swinging breasts painted with sacred patterns. Applied with ceremonial song, the patterns transformed the women so they were at once themselves, the embodiment of the ancestors and part of Country, sewn into its very fabric. Accompanied by a chorus of song, they danced with gusto the stories of a priapic old ancestral man, all penis and not much brain, chasing young women through the desert, tripping over himself with ardour—each gesture gouging features into the land. It was as much as the women could do to maintain the storyline, rolling about in mirth, wiping eyes wet with laughter. The story runs deep, though, and the women will not wash until the ochre is worn off their skin or bound to it. While the original inhabitants of Country have given it a human presence, Country, in turn, has given its people a voice and a presence that doesn't stop at the skin but continues out into the land.

Country—its land, sky, wind, animals, plants, earth, rivers, sea, rain, and the way they intertwine—provides the logos of Law. The Law is first and foremost place-based. The shape and content of each person's Country are constant and unambiguously provide a matrix for ecological and political stability,[8] seamlessly binding an intimate knowledge of nature into all social and economic relations and ceremony, and placing obligations on all of creation, not just people. Living things communicate by their sounds, smells, actions and non-actions.

Country calls to people, beckoning certain folk as food comes into plenty. Every sound or movement is followed with rapt attention. In this cosmology, all creatures have a rich social life with family quarrels, disobedient children, and unfaithful husbands and wives. But all follow the Law because they are all manifestations of the same essential being, as are rivers, mountains, planets and stars.

Totems weld together a person and their specific human and non-human groups. A person can have more than one totem, ensuring that ties bind different countries in familial relationships. Totems do not need to be of direct benefit to humans. Certain plants of little use to humans provide food for birds, turtles, fish and flying foxes. Other fruiting plants are good for humans and for turkeys and dingoes. Pollen provides food for bees so they can make sugarbag (honey). A certain fig is good for making implements and is eaten by turtles. That fig also announces the time when turtles are fat. Ties are recursive, and ecosystems flourish through looped and entangled relationships. Knowledge is deeply ecological and highly localised, yet it is woven across the continent like a strand in a dillybag. Deborah Bird Rose explains it like this: 'Events tell what is going on and call for action. One call leads to another, so that action is both a response and message . . . Country tells what is happening; it announces its own patterned eventfulness and invites engagement.'[9]

When people in western Cape York went back to Country after the disease and hideous killing times of early last century, they found that the land, which had not seen humans for many years, had 'gone wild'—the scrub thick, the grass long and unburnt, footpaths closed up, and open ridge tops grown over. People approached the intensely alive Country nervously and with extreme caution, as forces angered by the lack of care might strike: a snake bite, spider bite, crocodile attack or tree fall could be revenge for neglect. To calm the feisty presences and make the area ritually safe, the recognisable underarm smell of a senior Traditional Custodian was given as he called out to ancestors and spirits.[10]

Aboriginal landscape awareness is drenched in religious sensibility, but equally the Dreaming is saturated with environmental consciousness. Theology and ecology are fused[11] as people work in tandem with the ancestors. So closely acquainted with the habits and mannerisms of wild creatures are they that in the evening, when the anecdotes of the day are related, the people seem to become each particular animal through their clever miming and imitation of sounds: 'the dull-witted emu, the great stupid bird twisting its neck in quick stilted movements as it tries to assess the significance of some unfamiliar sound or movement,' wrote Edward Docker in western New South Wales.

> Then a hunter leaps from behind a tree, and the emu seems to gather up its skirts and run helter-skelter across the plain, with the delighted audience shouting derisive jokes after it, and very likely pretending to identify it with some aged relative whom it is safe to mock. If the performance is at all clumsy or unconvincing the audience laughs and boos the actors off the stage.[12]

In the desert, one observer wrote of one ceremony that, even though there were no body decorations or painted shields, was an unforgettable example of beauty in movement and artistry in acting and mime.[13]

The relationship of people to Country is uniquely expressed in ceremony, dance and song. Deborah Bird Rose wrote that when you dance, 'you are dancing the earth, and the earth is dancing you, and so perhaps you are motion, a sound, a wave of connection'[14] allowing the Dreaming to unfold and become animate.

In the jungles of the Wet Tropics, author Ion Idriess wrote of a ceremony held on a still night in a grassy pocket hidden within the depths of the rainforest. Earlier, the men had painted themselves with pipe clay and used blood to adhere the immaculate white down feathers of the sulphur-crested cockatoo. Some had down feathers only on certain parts of the body, on one leg and not the other, while others had circular bands and still others had stripes. They were aids to illusion. Waiting, Idriess watched the full moon rise high enough to pour light into the pocket, silvering the people, the waiting trees and the ceremonial ground. Women sat with their legs folded under them, forming a semicircle.

> As one, their clasped hands rose and came sharply down between their thighs, and the human drum in a startling crash reverberated throughout the clearing again and again and again. From behind them, within the black jungle, a sudden roar of voices was sustained in an animal savagery as out glided ghosts—some, I could swear, were floating—ghosts that were living skeletons, ghosts all white and seeming almost shapeless, others with but one ghostly arm, one ghostly leg, others all white but headless, others all body but legless. The howl in those voices, which now all merged into one, fairly made my hair stand on end as they came gliding past the human drum that now was fairly crashing with sound as the oncoming phalanx, with stamping feet, added thunder to the sound. I actually felt the earth vibrate as with one shuddering roar they came

bounding straight toward us, waving spears and clubs and those mighty swords.[15]

Correct performance of ceremony as well as attendant and obligatory land management calls to ecosystems. Poor management is met with punishment: blinding if a rainforest patch is mistakenly burned, or collapsed numbers of yam bulbs if the tops are not put back in the soil.[16]

And land, in turn, calls to be managed so it can be shepherded safely through shifts and changes.

# CHAPTER TWO

# The getting
# of wisdom

A rainforest is not just timber and trees, an annoying cloying clumping of jumble and tangle, endless and meaningless. Glances are intimate, complex, incomprehensible: shattered sunlight and torn sky, vertical rivers of rain, unearthly mist, trees flowered with stars on clear nights, and rivers of sky above sonorous streams.

A rainforest's silences are a polyphony of subtle, ceaseless, scintillating noises. A humid wet chaos of activity. The scented wind is alive, complex, mouldy and sweet, earthy, varietal: an evolutionary vintage. Distillations of ancient plants embody the living history of living plants. Three hundred million years ago—and beyond—a bridge formed between reality and magic, potent with memories that go deep in time, like dreams hardwired into grey matter.

———

At the scene of our ambush, things were heating up. The tropical sun of Far North Queensland burned away the mist to reveal the patchwork landscape of the Atherton Tablelands over which Thiaki broods, clinging to foothills at 1000 metres. Seeing our stupefied expressions—and no doubt thinking of the sale—Barry said, 'Want to go have a look?' We jumped into the back of his ute and held on to the tray. His place was called 'Ups and Downs', and that's exactly what it was.

Much of the land like this was 'up to shit' according to a settler of one of the last areas opened for clearing in 1954, 18 kilometres from Thiaki. He had to hang on to tree trunks as he was brush-hooking around steep hills, all while it was raining like hell. The same settler wrote in 1979 that you needed 'a fistful of dollars, a big heart, no brains and a bloody big umbrella'[1] to farm this country. He wished he'd walked off it in 1954. The soil proved marginal for farming and, once cleared and burned, it reverted to a maelstrom of weedy lantana, bracken and tobacco bush when the farm was abandoned. All but four per cent of the 80,000 hectares of rainforest on the Atherton Tablelands were razed after the bomb euphemistically called the yeoman's idyll resulted in close, cleared settlement, which only dissipated with the World Heritage listing of the Wet Tropics in 1988.[2]

Thrilled with the place, we bought Thiaki in 2005: our 180 hectares comprised 130 hectares of logged but intact forest and a steep 50-hectare paddock. With value measured in cows, we would struggle to make a living from Thiaki even though rainforest was among the most biologically valuable pieces of land in the country. At best we would receive $5000 a year agisting Barry's cattle in the precipitous paddock—which was too steep for crops—where the cattle had trampled narrow terraces, like linear paddies, on which to walk along contours.

The Wet Tropics covers a minuscule 0.3 per cent of Australia's landscape but supports more species diversity than any other place on the continent. Thiaki is at the western extremity of the World Heritage Area and about 1 kilometre from one of its convoluted boundaries. The nearest town is Malanda, 15 kilometres to the north by road.

Herberton lies a similar distance to the west in a straight line across an arm of the Great Dividing Range called Herberton Range.

While it hadn't been easy to turn lush tropical rainforest into paddocks—some individual settlers worked at it for decades—how could you turn paddocks back into rainforest? Local efforts on the Atherton Tablelands to restore degraded land back to rainforest were heroic but ad hoc. Community efforts relied on meagre government environment funding and focused on planting dense stands of native seedlings to form an instant closed canopy. It was expensive—about $60,000 a hectare—so only a few tens of hectares were planted. No matter how carefully the plants were placed (such as along creek lines), this expensive approach could not underpin the landscape-scale change that was desperately needed to restore the country to any semblance of its former glory.

So, with colleagues from several universities, and some of our own funds as leverage, we pitched an idea to the Australian Research Council during one of their annual rounds. We were jubilant when, a year later, in 2008, we heard that our application was successful. We would aim to tease out restoration methods that maximised benefits to conservation by working with the rainforest and allowing it some agency, while at the same time providing an income through carbon sequestration. We would set up one of the few experimental demonstration reforestation sites in the world to examine different approaches and associated costs.

So far so good. Work commenced apace to design the reforestation research plan for the foundations of a brand-new rainforest that uses different mixes and numbers of native species with different planting densities. There would also be grass-management systems with no plantings to see if the rainforest would kickstart restoration itself. On 28 January 2011, a bunch of academics and a crack team of planters with dirt under their fingernails and dreadlocks like flowing lianas, assembled to build the foundations of a rainforest with 30,000 plants, and to create the conditions for a phoenix to rise once more from the ashes.

The date in January was important. Noel had looked back as far as rainfall records went to see when rain would be a given and the ground would be guaranteed to be wet enough to accept our seedlings. But as the week progressed into February, the sky became clearer and the weather hotter. Strange streaks appeared across the blue. It was ominously still, and a silver gull fled from the coast over the ranges inland. It was a warning. Something was sucking away our weather. Six days of planting later, and a day after the last seedling went into dry ground, Severe Tropical Cyclone Yasi—category 5, and one of biggest cyclones ever to cross the coast—roared across the Coral Sea and over Thiaki.

———

I touch the fleshy liver-shaped lobe of a plant clinging to a tree trunk in the moist shadows of Thiaki. Its single cells brush against my skin cells, and I find myself communing with a liverwort. A palette of liverworts and mosses is smeared across the trees that are set against a half-light canvas. It is a cool place to hang out, where clumps of mosses droop down from branches in scruffy beards. One of the largest mosses in today's world—with a stem growing up to 20 centimetres long—*Dawsonia polytrichoides* carpets sun-spilt gaps on the forest floor.

Liverworts and mosses are known informally as bryophytes, or moss plants. They lack rigid vessels to transport water and nutrients, instead taking these in by osmosis, and so they are restricted to moist areas. Instead of roots, they have fine chain-like filaments with single-cell links that anchor the plant. Bryophytes strip water from passing clouds, harvesting up to 40 per cent more water than is detected in rain gauges. In dry months, 70 per cent of Thiaki's water will come from cloud stripping.[3]

I stop caressing the liverwort, not at all fooled by its humble appearance. While it clutches the tree tenuously, it is, in fact, the tree that is metaphorically standing on the shoulders of this tiny giant. This is because liverworts were among the ancestors of all land plants that began as a smudge of cells, a spark of terrestrial life, in the glutinous muck that fringed ancient swamps, rivers and pools. Captured fully formed as fossils in 480-million-year-old rock,[4] these ancestral plants evolved much earlier in the primordial soup of 600 million years ago on the pre-Pangaean continent of Pannotia when all of the continents of Earth were clumped south of the equator.

Little more than jumped-up green algae in the Ediacaran ocean, liverworts crept onto a bare, mostly lifeless landscape with its fringe of moist microbial soil crust and micro-arthropods, beginning an experiment of cohabitation. Sex, seeds, erect woody tissue, stomata for breathing, and vessels for transporting nutrients from deep roots were all distant imaginings of this collection of cells. Liverworts passively absorbed moisture directly into cells and reproduced through passionless spores. Their food and energy came from the sun in the miraculous but unstoppable process of photosynthesis, which emerged soon after life itself. Some cells had lignin but, being self-sufficient in their moist patches that allowed sperm to swim about freely in search of eggs, there was no need to become erect. Nevertheless, the familiar characteristics of photosynthesis, cell walls, sperm and eggs mark bryophytes as the direct ancestors of the towering forests of Earth, and of Thiaki.

It is quite a story—cell-sized micro-beings creeping slowly and inexorably onto barren land, forming a biological crust that morphed over unimaginable periods of time into towering forests that shaped the land—and to my mind it is not at all a far cry from Aboriginal cosmology, which tells of a desolate land made lively by ancestors.

Today, there are about 20,000 species of bryophytes in the world, and they grow on forest logs, in acid bogs, on tree trunks, near swift-flowing streams—and on my chairs, windows and outdoor speakers. Bryophytes remain one of the largest groups of land plants. There are as many liverworts in Thiaki as ferns and conifers.

Having made the march onto land, long before terrestrial verte-brates lumbered from the sea, liverwort-like plants dominated it for 50 million years,[5] brushing the contours of the virgin landscape with mossy greens.

———

Who can know the generational struggles of bryophytes? They must surely have been heroic because they lived through the 'Big Five' mass-extinction events, including the first one around 445 million years ago where nearly 85 per cent of species on Earth were wiped out. This extinction turned an ocean teeming with life into one where only simple ecosystems remained. The culprits for the extinction are hazy so far back in the mists of time, but a massive glaciation is implicated. It was triggered by the colli-sion of Africa and South America: they smashed together into a vast mountainous supercontinent, which then drifted over the South Pole.

Plants themselves are suspect. At the time, they had begun their terrestrial takeover under a deadly sky of hydrogen, nitrogen, water vapour and carbon dioxide, with very little oxygen. Early plants may have metabolised the heat-creating carbon dioxide and tipped Earth into a cooling vortex. Being persistent, they may also have worked their way into rock, triggering the release of fertilising minerals such as phosphorus that catalysed bursts of algal blooms. In turn, these sucked up what little oxygen there was, dooming most animal life to extinction.[6]

In this hazardous Lilliputian land where club-moss canopies were only 10 centimetres high, certain bryophytes nevertheless survived and evolved. Lignin was put to use to harden vessels so they could transport fluids and nutrients, and to brace the skyward-soaring plant as it released its spores into the wind. The nutrient-bearing phloem became the pathway for electrical communication. As in animals, plants' nervous systems are based on action potentials transmitted along the vascular conduits stretching throughout the body.[7]

Fern-like branching structures were refashioned into leaves. Dropping into mud, some of these became the first-ever fossilised impressions of ancestral land plants and are found in Victoria.[8] Around the world, similar fossils dating to 400 million years ago, during the Devonian, show an influx of land plants that already had well-developed internal networks of vessels. The club moss *Baragwanathia*, only the size of a small shrub, was a giant of its time. Its strong vascular network allowed it to evolve into tree-sized plants[9] and the first forests. Club mosses still survive to this day as the pendulous tassel ferns in the Wet Tropics forest. Thiaki harbours at least two species—*Phlegmariurus marsupiiformis* and *P. filiformis*—which grow on trees and rocks in the shadows of the deep valleys, or fall in glorious plaits from staghorn epiphytes.

It is the wisdom of plants to form connections with the wind, the earth and living things, connecting any point to any other point in space and time. European philosophers have even used the rhizome[10] as a metaphor for establishing multidimensional connections with all life. While a rhizome is an underground part of a plant, it not only roots but also shoots up new stems from its nodes. Stems of ferns are called rhizomes. Rhizomes can also store nutrients in underground tubers. A scrap of rhizome is tenacious and will strike to form a new plant.

The evolution of plants and the creation of the atmosphere are intimately linked. Plants grew and died, organic material became buried, and oxygen was produced as a by-product.[11] Like micro-bulldozers, early roots disrupted the soil grain by grain, releasing minerals that interacted with, and locked up, much of the atmospheric carbon dioxide. (Today, enhanced rock weathering is being explored as a strategy to accelerate carbon sequestration; it involves spreading crushed rock over croplands to help remove carbon dioxide from the atmosphere.[12]) Over aeons, these chemical processes raised the oxygen levels in the atmosphere and fuelled the evolution of life on Earth.

Bryophytes are known as primitive plants, but plants that have survived since the time animal life comprised the likes of jellyfish and sea pens, and the atmosphere was one of poisonous gasses, know a trick or two. Long before plant roots evolved, bryophytes had already formed the first carbon-exchange system in the world— called a mycorrhiza—with semi-aquatic fungi. The cellular root-like hyphae of fungi couriered hard-won nutrients from hostile rock directly into the bryophytes' cells in exchange for organic carbon from photosynthesis.[13] The very evolution of plants is possible only because of this ancient kinship among living things.

Today, this relationship between fungi and plants is found in 85 per cent of plants from the tropics to the poles. The relationship can take several forms, but the dominant one is where the fungi grow inside the plant in the space between cells. From here, the fungi penetrate the cells and form branched structures called arbuscules, from the Latin word *arbuscula*, which means 'little tree'. Roots themselves probably evolved as structures to facilitate this interaction between plants and the arbuscular mycorrhizal fungi.

———

I looked at our paddock with its sun-blasted soils, leached of nutrients and monopolised by grass. It was in a state of arrested evolution—an expression that surely captures the essence of the Anthropocene. The question was how to kickstart a rainforest again.

We designed our rainforest restoration project with a team of scientists, local practitioners and nursery managers. A series of reforestation experiments would test what species to use, how far apart to plant them, how to get them into the ground cost-effectively, and which weed control to use. Twenty-four native species were selected based on their availability and 'functional traits': things such as how tall they grow, how dense their wood is, what sort of fruit they have and how the seeds are dispersed, how the flowers are pollinated, and how they get their nutrients. Experimental plots of monocultures, plots of six mixed species, and plots of 24 mixed species were planted. The idea was to see what method would produce the most cost-effective but still self-sustaining forest.

*Flindersia brayleyana* (Queensland maple) was chosen as the monoculture because of its use in restoration and as a plantation species. We could have used another frequently grown native, *Cardwellia sublimis* (northern silky oak), but in plantations elsewhere it was not as successful as *F. brayleyana*. The idea of using a monoculture was to see if it would form a living framework through and onto which a rainforest would grow. Birds, for instance, perch on branches and disperse seeds on the soil below with their droppings. And the shade from the fast-growing trees would protect the new rainforest as it found its feet and outgrew the framework.

From our home, you can look down the main valley and see a patchwork of young trees growing up the steep slopes. Miraculously, Severe Tropical Cyclone Yasi's fury barely touched the newly planted seedlings. Quite ironically, it was the irksome grass that saved the day. We had left grass between the weed-sprayed rows of young seedlings, to reduce costly pesticide use and protect the soil microbiome, and the grass had blown over the seedlings, shielding them from the 250 kilometre per hour winds that raked the landscape.

Three years later, in 2014, we noticed that the *F. brayleyana* trees growing in a plot on a particularly steep slope were not looking well. Where the slope was steepest, the young trees were spindly, less than a metre tall, and they looked jaundiced. At the bottom of the slope, where it flattened out in the valley, they were lush, green and growing fast—already 6 metres high. It looked like we had two completely different species instead of one species with sickly and healthy populations. In the adjacent plot, with a six species mix including *F. brayleyana* and *C. sublimis*, the *F. brayleyana* trees were again spindly and sickly on the slopes, but the *C. sublimis* trees were robust and seemingly bounding up the slope with no difference in size. One day, with a neighbour, we were scratching our heads about the whole thing when she said with typical country wisdom that all the goodness had been washed down to the valley. And she turned out to be right.

With two James Cook University friends, experts in plant physiology, we decided to investigate. We spent a weekend on our hands and knees, digging up and bagging multiple samples of the soil and leaves from the two plots, and then putting these through a barrage of measurements and tests. We discovered that one of the characteristics of *F. brayleyana* is its reliance on arbuscular mycorrhizal fungi.[14] It seems that our overgrazed, slope-washed soils are too depleted even for these highly evolved relationships. Phosphorus, in particular, is in short supply except where it is washed downslope and becomes a pool of nutrients for the mycorrhizal fungi that energetically pump them into the healthy *F. brayleyana* trees.

The question is what happens to the *C. sublimis* trees, which seem to ignore the nutrient deficiency?

Life loves to innovate, and it does this with unparalleled brilliance: it will think up, concoct, invent, reinvent, destroy and resurrect. It relishes trialling alternative solutions to a problem. So, instead of using mycorrhizal fungi, *C. sublimis* has taken the plant roots and fashioned them into something new. When the tree detects that its nutrient levels have fallen below a certain critical point,[15] super rootlets erupt from the main roots in dense clusters, growing quickly in a matter of days. These burst and release organic acids that mobilise recalcitrant nutrients, such as phosphate and iron, which are tightly bound in the soil. Once mobile, these nutrients are quickly captured by the plant.

There are downsides to this strategy, the most obvious being the inherent physical cost of regularly growing clusters of dense roots (like suddenly growing a new arm when needed). And, unlike mycorrhizal fungi, cluster roots do not take advantage of very high nutrient availability, which explains why the *F. brayleyana* trees, charged like a battery with arbuscular mycorrhizal fungi, overtop the *C. sublimis* trees on the lower slopes where nutrients have pooled.

———

While the club mosses and ferns were experimenting with life on land, the oxygen vented from their exuberantly evolving forests spawned a revolution in the sea. Animal evolution, which had been stunted by hypoxia, was suddenly fuelled by oxygen and nutrient-rich run-off, and it flourished into the age of fishes around 400 million years ago. Marine vertebrates and predatory fish[16] grew to over 2 metres in length. And the first tetrapods began to leave fossilised tracks of their four fin-like feet as they shoved them into the mud of a 395-million-year-old swampy shore.[17]

One of these tetrapods became a labyrinthodont[18]—like a huge axolotl but with a bigger head and hundreds of sharp maze-like teeth. Labyrinthodonts could grow up to 5 metres in length, had primitive lungs (like today's lungfish) and were the dominant verte-brate animal from 350 to 210 million years ago. They lumbered about on stumpy legs near water, sinking into the moist soil and litter already packed with invertebrate mites and ticks strikingly similar to those of today.

‘He's had a heart attack,’ said the doctor. Noel had developed severe pain across his upper back and had had it for a couple of days. We tried massage, hot showers and packs, but the pain worsened. Late on the third afternoon, I drove him to the hospital and the emergency doctors immediately diagnosed a heart attack. ‘I think you need to look at the tick bite on his shoulder,’ I said. ‘He's not having a heart attack.’ The doctors at the small country hospital at Atherton looked at me quizzically and insisted, looking at my middle-aged husband, that he'd had a heart attack. His heart was pumping out specific cardiac distress hormones, indicating it was in a bad way, so a decision was made to rush him by ambulance to the large Cairns Base Hospital.

I raced home in disbelief. Noel is lean and fit, and he has perfect blood pressure. We eat well and exercise daily. I didn't believe he was having a heart attack. I grabbed some clothes and headed to Cairns, an hour and a half away, unsuccessfully resisting the urge to speed. In Cairns, the doctors had reduced his pain. Once again, I pointed out the strange-looking tick bite with its halo of red necrotising skin that had been there for about ten days. I was told, quite firmly, that he'd had a heart attack. It was late, and I was advised to go home. They'd ring if his condition changed.

As I drove up the steep Kuranda Range to the Atherton Tablelands, the phone rang. I was told to turn around and drive carefully back to the hospital. 'His condition has changed,' came the soothing voice. What the person really meant was to concentrate on the road but get there before Noel dies. He had been moved to the private hospital when they'd realised we had private cover. But his condition had deteriorated, and he was having an angiography (where dye is run through the heart) to see what the problem was. A couple of doctors and technicians were crouching over computer screens in a dimly lit room. I could see Noel through the large glass window, lying flat. I wasn't sure if he was alive.

I was told to go into an adjacent small waiting room. It was empty. All I could hear was the maddening hum of the air conditioning. There was a picture on the wall that I spent what seemed like hours looking at. I still can't remember what it was of. The door opened, and a doctor walked in. Time stopped. He smiled gently and said, 'He's not having a heart attack.' I wanted to punch him in the nose.

I was taken into the room to see the computer images that showed Noel's heart pumping rhythmically. 'I'd be very happy with a heart like that,' said one of the technicians cheerily. Nevertheless, the pericardium of Noel's heart had inexplicably expanded, sending signals that indicated a severe heart attack—and that was not good. The smiling doctor said that they had asked a specialist in tropical diseases to look at the tick bite on Noel's shoulder.

The next day, the tropical-disease specialist became visibly excited when he saw the tick bite. The halo around the bite, called an eschar, was diagnostic of a number of diseases caused by bacteria (such as *Rickettsia*) that are carried by ticks. It wasn't that common, and it was rare that a patient had Noel's heart symptoms. The doctor took a biopsy, drilling out a neat fleshy core ('this won't hurt much'), and found that Noel indeed was infected with *Rickettsia*. He was immediately started on a course of antibiotics, and a couple of days later he was out of hospital and on beta-blockers while his heart recovered.

Together with the tropical-disease specialist, local veterinarians and field ecologist colleagues, we spent the next few years sending ticks from tree-kangaroos, northern bettongs, bandicoots, our cattle dog and us to the Australian Rickettsial Reference Laboratory—which meant that I had to leave any ticks I found on me embedded in my body until they could be carefully tweezered out and dropped into a vial to send live through Australia Post. Of these peripatetic ticks, six per cent carried *Rickettsia* bacteria—including *Ixodes holocyclus*, the paralysis tick, which nearly killed Noel.

Back in the early days of life, the ancestors of *Rickettsia* were free-living bacteria in sea water.[19] The explosion of multicellular beings in the soup of life enabled them to bump into and serendipitously merge with other cells. Here *Rickettsia*, ever the opportunist, ditched its own DNA and began manipulating that of its host. The bacteria jumped from host to host and found its way into the first terrestrial arthropods, the mites and proto-ticks, which were living in the wet litter around shorelines dotted with abundant, sluggish labyrinthodonts full of nutritious blood. All it took was for a proto-tick to discover a way to extract this ready-made meal and abandon the competitive life of soil litter.[20]

Today, *Rickettsia* still lives in the saliva of Thiaki's mites and ticks. Our rainforest seedlings adjacent to a sea of pasture provide such ideal conditions for tick-infested bandicoots that we seem to be tick central. Indeed, so numerous are our ticks that each year an employee of the manufacturer of a tick-protection treatment for pets comes to Thiaki to harvest ticks. Armed with a blanket, he literally drags it across the paddocks and new plantation plots like a prawn trawler dragging the seabed, then picks off the hundreds of ticks that become stuck to the blanket fibre.

———

Six hundred million years ago, when *Rickettsia* were free-living bacteria, the land that would become Australia began a slow waltz across the globe. It was originally in the northern hemisphere, jutting out of the eastern coastline of a vast supercontinent that embraced India and Antarctica; bits of China poked out nearby as a vast northern finger. Australia was tipped on its side, at a right angle to its present-day position, with the Pilbara coast—its most southerly point—resting on the equator, and its northern coastline running from the Flinders Ranges through Central Australia and north towards Arnhem Land.[21] Much of what is now the eastern part of Australia had yet to emerge from the Tethys Ocean, and therefore Thiaki lay under open water.

As part of a flotilla of continental fragments, Australia drifted south on its tectonic plate across the equator, doing a slow clockwise pirouette. The warm equatorial environments on the edges of quiet embayments and swamps supported luxuriant club-moss forests that were making a stand against the glacial cataclysm building at the South Pole by the collision of Africa and South America.[22]

As Australia continued to slip further into the southern hemisphere, its eastern margin erupted into an 1800-kilometre arc of oceanic volcanoes from near Sydney to northern Queensland, extending inland of Townsville and Cairns and up to the Torres Strait,[23] as ocean crust from the pre-Pacific Panthalassa Ocean was driven under the continent. Earth's continents were in a colossal collision that would form the giant super continent of Pangaea, which would straddle the planet from pole to pole. It wasn't a friendly union. Continental-scale volcanic activity in Siberia thrust up gargantuan magma flows that filled massive 800-kilometre long and 450-kilometre wide valleys in a lava pile 7 kilometres thick;[24] the paroxysms blasted out toxic gases that poisoned everything.

Around 360 million years ago, the island-arc system alongside eastern Australia began piling up, sending Thiaki soaring thousands of metres skywards. Australia did a pirouette of 180 degrees, so that it now lay on its other side: the east coast formed the southern extremity of Australia, lashed by polar seas, with much of today's southern Australia well inside the Antarctic Circle and attached to Antarctica. The tip of today's Pilbara formed the north coast, and it sat at about the latitude of today's southern Tasmania,[25] washed by the Tethys Ocean.

While massive ice sheets sheared life from the southern half of Australia, on the fringes of the ice sheets new plants were evolving, resolute and determined to adapt to cold climates, experimenting— and then thriving in the new conditions. These plants, loosely called seed ferns, broke away from a dependence on free-swimming sperm that needed the medium of water for reproduction. Instead, they innovated a protective coat for the sperm, called pollen, and they enclosed the female eggs in little cups that developed into seeds when fertilised.

In what was one of the most significant events in the evolution of land-based vegetation,[26] forests began appearing on the frontier of continental dry land around 300 million years ago. Thiaki lay under a polar twilight on the edge of vast plains that were carved by mean-dering rivers and blanketed in mats of vegetation, which became peat bogs and, later, coal.

Giant dragonflies had already mastered flight. Millipedes, centipedes, scorpions, spiders and cockroaches buzzed, droned and scurried about. The moist ground was laced with tracks and footprints from a multitude of bizarre and wonderful reptile-like creatures. Some were squat and small, others were gigantic, still others had a dorsal sail to regulate temperature. Most of these creatures would die out, never to be seen again. A tiny few would evolve into dinosaurs, birds and mammals.

Club mosses grew like small palms in today's mangrove niche, while horsetails congregated on the muddy margins like rushes. Deciduous *Glossopteris* seed ferns grew in dense woody clumps in cold swampy bogs, dropping piles of tongue-shaped leaves in anticipation of an icy winter.

———

In Antarctica in 1912, Robert Scott, leader of the fateful *Terra Nova* expedition to the South Pole, and two expeditioners were found frozen to death in their tent. Recovered from their nearby sled was a fossil of *Glossopteris* that had been picked up on the Beardmore Glacier on the return journey. Overshadowed by the deaths of Scott and his party, the scientific contribution of the enterprise (known formally as the British Antarctic Expedition) was significant. Twelve scientists participated in the expedition, and the *Terra Nova* returned to England with 2100 plants, animals and fossils.[27] The fossil of *Glossopteris* was a surprise to the geologists of the expedition, who were not expecting to find fossil forests on Antarctica.

Even more perplexing to them was that the fossil was almost identical to ones found in India and in the Sydney–Bowen Basin, a region south of Thiaki considered by today's geologists as a single geographic province with similar species of plants.[28] This was the first piece of evidence that proved Antarctica was part of a great southern-hemisphere continent called Gondwanaland—now commonly called Gondwana—named after the Indian district, Gondwana, in which the first *Glossopteris* fossils were found. In the same year as Scott's death, Alfred Wegener announced his theory of continental drift using *Glossopteris* as a key example.[29]

*Glossopteris* forests were the signature forests of Australia at the time, as eucalypts and acacias are today, and featured seed ferns shaped like Christmas trees that taper towards the top. A species-rich genus with members growing up to 30 metres tall, *Glossopteris* was widespread in mire environments across Gondwana. Thick seams of coal in Australia's basins are the evidence of the profusion of these early forests. Thiaki was clothed in a voluptuous rainforest of ferns and *Glossopteris* seed ferns. Across the horizon, the first conifers and early cycads—evolved from an unknown seed-fern ancestor—advanced over the dry hillsides. The vividly coloured fruits on the newly evolved cycads were a bright red beacon to co-evolving animals.

———

Cycads have been called plant evolution's Rosetta Stone.[30] Just as the message carved on this slab of stone allowed archaeologists to read the Egyptian hieroglyphs, the features of cycads allowed the rest of plant evolution to be understood.

Evolved from one of the earliest seed plants, cycads are rare survivors from this ancient time 300 million years ago. They track the path of the emergence of modern plants' fruits and seeds.[31] In cycads, the male sperm cell, packaged into pollen, enters the female cone, a structure of freshly modified special leaves that enfold the egg. The egg is buried deep within several nutritious layers laid down to feed the embryo. The sperm cell burrows slowly down to the egg, a process that can take months. Near the egg is a small drop of water secreted by the cone. There, the sperm swims in a miniature re-enactment of the journey through primordial seas of its algal ancestors.[32]

Whatever ate the cycads back then must have had a legendary constitution because the seeds were (and still are) deadly. They contain the neurotoxin methylazoxymethanol acetate, which damages the brain and liver, and causes cancer. Symptoms of cycad poisoning in animals include lack of muscle control, memory loss, altered speech, confusion and disorientation (similar to Alzheimer's disease), convulsions and neurodegeneration.[33]

About 315 million years ago, a new type of four-legged animal emerged from the swamps—the amniote. While it didn't look very different from the other amphibians and early lizards, it started a trend that became all the rage. It laid eggs not in water, but on land, which was a convenient habit if you were going to make a living on land. A few million years later, the amniotes split into sauropsids—giving rise to all reptiles and birds—and synapsids, which would evolve into mammals.

While early synapsids were carnivorous like their reptile and amphibian cousins, unlike them they began to evolve into both carnivores and herbivores. Some carnivores grew to the size of bears. And some of the herbivores weighed up to 2 tonnes, big enough to house their enormous fermenting stomachs.[34] The synapsid herbivores were likely the first animals to get around in great lumbering ancestral herds to reduce their chances of being eaten by the fierce carnivorous synapsids—a pattern woven into the fabric of life to this day.

The early herbivores didn't have the teeth to chew vegetation. Instead, they swallowed the enticing masses of red cycads[35] in huge gulps that didn't disturb the toxic seeds buried deeply within the edible fleshy fruits. The evidence for these table manners came much later when a dinosaur fossil was recovered in 2009, from 180-million-year-old South American sediments, with a belly full of intact cycad fruits. This 5-metre-long, 2-metre-high herbivorous dinosaur had wandered about on two strong hind legs, gobbling down cycad fruits without chewing.[36] This way, the seeds made it through the digestive tract of the animal unscathed, to be dropped across the landscape ready to germinate and prosper—the poisonous chemicals were possibly used to protect the new seedling from rapidly evolving insects. Shortly after its meal, the gluttonous dinosaur became bogged in deltaic muds, leaving its skeleton and stomach contents to posterity.

Perhaps the strangest of all creatures to eat cycads are the comparatively delicate synapsids called humans.

'I ate five or six of them . . . but after about three hours I and five others who had eaten of these Fruits, began to vomit so violently that there was hardly any difference between us and death.' In 1697, almost a century before British colonisation of Australia, Dutch explorer Willem de Vlamingh's crew experienced ill effects from eating *Macrozamia* nuts that they had found scattered around Aboriginal camp sites along a river peppered with swans (now Perth) and had therefore assumed to be edible.[37] One of Vlamingh's crew described the fruits as being scarlet and containing a nut not unlike the chestnut and not unappetising. Unfortunately, eating the nuts caused a form of vertigo that resembled madness, as the affected mariners crawled on the ground and made senseless gestures for two days. The toxic nuts had, in fact, been discarded by the local people in favour of the fruity flesh, which had been buried so it would ferment, enhancing the flavour and nutritional value.

Across the other side of the continent, fire and water were wielded to make the toxic nuts not only edible but also a staple. So important were cycads to Kuku Yalanji rainforest peoples, who lived a short distance north of Thiaki, that in the Dreamtime an ancestor created woman from man not only for sex but also so she could detoxify the nuts and make them delicious.[38]

Joseph Banks, botanist on the *Endeavour* in 1770, noted that the leaves of the cycads, which he called a type of palm, resembled a fern. This was an astute observation because the leaves of cycads still carry their ancestral fern-like form, the leaflets unfolding along the vein exactly like a fern. Banks noted that they bore a plentiful crop of nuts 'about the size of a large chestnut and rounder'. The hulls were found plentifully near people's fires, and some sailors being confident that the nuts were therefore edible once again tried some. 'They were deterred from a second experiment by a hearty fit of vomiting and purging,' Banks noted coolly. The ship carried pigs and, being short of food, they ate the cycad nuts even more heartily than the sailors. After a week, all of the pigs became extremely ill and two died. Banks concluded 'it is probable that these people have some method of preparing them by which their poisonous quality is destroyed'.[39]

The process of detoxification is complicated and only women know it intimately. It involves the gathering of carefully selected nuts, which are roasted and then grated by groups of women sitting on rocks, laughing and talking in the sun. The resulting flour is sifted through a palm-fibre dillybag; the fine particles thrown into the water for the turtles and fish, and the coarse particles put into an intricate grass dillybag that has been carefully folded into a basin, like a colander, and placed near a stream. Leaves are carefully fashioned into a funnel that directs water through the basin to percolate for a day or more. The resulting paste is ready for eating but is often matured for a few days more to intensify the flavour. On the lower Tully River, to the south of Thiaki, the nuts are steamed and then grated using the edge of a sharp shell. This pulp is put into a dillybag where rushing water falls from a height to keep the mash strained and well stirred.[40]

Cycads grow in pyrophytic groves that, on the edge of the rainforest, are carefully husbanded with fire to produce respectable and predictable yields. It takes experience to wield fire like this. Taught by Elders and guided by ancestors, the women apply fire with scientific precision. It must be hot for the cycads, but cool for the nearby yams, and just a trickle around the edges of the rainforest.

# CHAPTER THREE

# The new world

A forest is retrospection and contemplation. Is it also grief? Forests remember. Trees carry in their rings the memory of rainfall, sunlight, storms, temperatures, fires: the climates and moods of millennia. Moments and episodes are revealed there if you look, long experience captured in a plant, an animal, a community, lifestyles and behaviours, adaptations, niches and places. The silences quickly overgrown with life.

———

One of my first experiences with our newly acquired Thiaki was to walk our boundaries. They had been cut in straight lines across Thiaki's voluptuous slopes and narrow valleys, and it took hours to walk the 4 kilometres or so, clinging to trees in the shifting soil and loose litter, avoiding stinging plants (the excruciating pain they cause—like getting an electric shock and being burned by acid all at once—can last months) and sharp lawyer vines (once they've got you, they don't let go) and wading through small, but sheer and slippery, rocky gorges and valleys with fast-flowing water.

In a far corner of the property, we discovered a little pool beneath a small waterfall shaded by dense vegetation. Sediment had washed into the pool from a neighbour's land—he was logging the last big trees from the property he was selling—so I couldn't see the bottom. But the temptation to soak in my newly acquired pool was too much. I stripped off and stepped in, naively expecting to land on rock. Instead, I sank to my knees in newly settled mud and then clambered out like an unevolved tetrapod to sit and dry out on a low mossy log. It was some christening!

Later that day, the plentiful curves, nooks and crannies of my body erupted in raised red bumps. 'Scrub itch,' said the Pembers, nodding knowingly over an afternoon XXXX Gold. 'Whatever you do, never sit on a rotting log.'

Scrub itch is caused by trombiculid mites, otherwise known as chiggers, and they are closely related to the mites that carry the bacteria that cause life-threatening scrub typhus. Surprisingly, given the pitiful state of knowledge of other invertebrates in the Wet Tropics (or anywhere, really), there's quite a bit known about this mite family. And that's because they bite humans—mostly accidentally—while they attack and feed on the first available animal. Virtually invisible, it is actually the 200-micrometre-sized larvae of the mites that bite. Once hatched from eggs in the soil, these tiny six-legged creatures climb low vegetation and logs, where they clump together in mite islands, awaiting a suitable host (or serendipitous white buttock). When one appears, they drop onto the host and crawl to thin areas of tissue to feed on lymph and skin. 'Ears, head, armpits, abdomen, genitalia, and the area around the tail are preferred.'[1] Yep.

When engorged, the larvae fall off and eventually turn into eight-legged, 1-millimetre-sized adult mites. Related to spiders, but distinct from them and more diverse, mites are predators, parasites, herbivores and detritivores, and they are found everywhere: on land, in the deep sea, in rivers, on high alpine peaks, in subterranean waters and at temperatures as high as 50 degrees Celsius in thermal springs; under lizard scales, in the cloacae of turtles, in the ears of moths, in the respiratory tubes of bees, in human skin pores, you name it—but perhaps not quite in the numbers they are found in soil. Worldwide, about 40,000 species of mites have been described, of perhaps a million species. Who knows? Amazingly, there can be up to one million mites in a square metre of soil, of which trombiculid mites are just one group.

If I bend down and grab a handful of soil, I might be holding 250,000 mites (in the depleted paddock, it would be a fraction of that number). Within this miraculous matrix, mites help to regulate microbial processes by breaking down litter and dead organic matter, and hunting microbes and micro-fauna and climbing onto big and small fauna. Since Thiaki emerged streaming from the sea, first as a swamp and then a rich *Glossopteris* forest, mites have been pivotal to the structure and function of the ecosystem, and to the formation of mycorrhizae. Up in the canopy, there is a symbiotic network of mites feeding on mosses, ferns, leaves, flowers, fruits, other arthropods and each other.[2] Ever since my tick-infested synapsid ancestor scuffled in the primordial litter, this community has been more important to Thiaki than I will ever be. The experience in the pool was just a slap-on-the-bottom reminder of that.

———

After the continents of Earth had coagulated into one massive landmass called Pangaea at the beginning of the Permian about 300 million years ago, you could walk from Siberia near the North Pole to Antarctica and Australia, which lay together with South America, Africa and India, drifting near the South Pole. In the swamps and rivers, monstrous-sized Permian amphibians called temnospondyls—labyrinthodonts that broadly resembled crocodiles but were more like salamanders—were flourishing. Australia's earliest record of tetrapods dates to about 330 million years ago. Bones of three different types of tetrapods, similar to those found across the globe at the time,[3] were discovered in central Queensland: two of these are amphibian ancestors, and the other is a distant amniote precursor to the synapsids.

By 300 million years ago, herds of herbivorous synapsids were roaming in vast numbers across a landscape unimpeded by intervening seas. These elephant-sized proto-mammals had tusks and turtle-like beaks instead of teeth. Stalking the proto-mammals were the Permian carnivorous synapsids, such as the gorgonopsid, which grew up to 4 metres in length and had huge powerful jaws and sabre teeth. The Panthalassa Ocean, wrapping around Earth on either side of Pangaea, was rich with marine life from giant scorpions to trilobites and sharks. Vast coral reefs abounded in the shallower seas, over which crept and swam sponges, ammonites, nautiloids, crinoids and brachiopods.

Australia lay in slanting polar light, but in the super-continental climate—where continents were distributed more or less evenly in a C-shaped arc around the equator—subtropical conditions prevailed for much of the year. There is nothing even remotely like this today. Temperatures might have plunged to –20 degrees Celsius in the dark polar winters. But summer temperatures were in the mid-twenties, which melted any accumulated snow and ice. Continental regions away from the poles were hot and arid.[4] Rainfall across the globe was mainly convective and fell over the oceans. Much of continental Pangaea was desert with coastal monsoonal rainfall.

Thiaki lay near the southern coast of Gondwana and close to the South Pole, which had a cool temperate climate with seasonally wet winters. To the north lay today's Western Australian coast, at about the same latitude as modern Tasmania, where the climate was tropical with a humid monsoonal summer. Across the globe it was generally warm. Life was spinning off into different forms, with the synapsids in the ascendency. In the world of flora, seed plants thrived in swampy conditions, and conifers had broken their tether to water and joined the cycads and hundreds of seed plants on the dry hillsides.

Conifers were clever. With a nervous system inherited from fern-like seed-plant ancestors,[5] they were as complex as the animals[6] lumbering about on the forest floor. Innovatively, they reduced their leaves into needles and smeared them with a waxy coating to reduce evaporation. Conifers hung on to their leaves, even during freezing seasons, without the need to wastefully discard them as the surrounding deciduous *Glossopteris* trees did. Even today, conifers only reluctantly discard leaves, bark or roots. Individual *Araucaria* leaves, such as those of the local *A. cunninghamii* (hoop pine), are known to survive on the tree for 25 years.[7]

Conifers experimented with their reproductive parts, folding some leaves into whorled hardened shoots that were cupped within protective leafy bracts. Onto these were fused simple pollen or ovule-bearing structures. Trial and error forged these into cones. Araucarian conifers, such as the *Agathis* genus (kauri), renovated their primitive single-celled vessels so that they could carry water from roots to leaves using smart valves to improve hydraulic efficiency and strength. So efficient are these valves that conifers became and remain to this day the tallest and longest-lived of plant species, the narrow vessels and smart valves doubling to protect the tree from freeze-thaw damage. On the Atherton Tablelands, the age of *A. microstachya* (bull kauri) can be measured in thousands of years[8]—the few remaining examples of these conifers on the cleared Atherton Tablelands may have been seedlings during the reign of King Egbert, the first monarch of Anglo-Saxon England (AD 827–839).

———

Thiaki was part of a mountainous landscape that featured broad lake-filled valleys interspersed with coral-encrusted marine embayments.[9] Fossils from a nearby coal bed reveal lush forests of *Glossopteris*, cycads, ferns and seed ferns.[10] Insects had quickly followed the plants to the new world and had formed specialised associations with them. At least 27 orders of insects had evolved, which is close to today's 29 orders.[11] (Each order can contain thousands of species: today, one order of insects, Coleoptera—the beetles—has around 350,000 species.)

The amphibians—from small lizard-like creatures to those the size of modern crocodiles—occupied an abundance of niches, forming a primeval trophic cascade: some ate the abundant fish of the swamps and lakes, some were mid-level predators, and others again were the higher predators preying on smaller animals. Somewhere in this thriving group were the ancestors of the archosaurs, those ruling reptiles who would give rise to dinosaurs, crocodiles and birds. Other reptiles, the captorhinomorphs, scurried and scampered among the trees.[12] One of these would evolve into the ancestor of lizards and snakes, and another into the ancestor of turtles. Where the waters were shallow, reptilian mesosaurs swam after crustaceans,[13] and the earliest of the frog and salamander ancestors were scrabbling through the mud and litter and around the algae-covered roots of swamp plants.

In this bizarre world of synapsids and reptiles, there was no birdsong but instead the chirping and stridulating of insects set against a background soundscape of roars, grunts, growls and perhaps the squeaks and plinks of frogs. Pelycosaurs roamed the plains of Pangaea bearing large sails held upright by stupendous spines, which may have been primitive solar panels. These synapsids were the first of Earth's megafauna—some animals measured up to 3.5 metres in length. Their fossils are found in North America, but in such a connected world they likely roamed across Thiaki's rich *Glossopteris* forests as well.

Pelycosaurs gave rise to a new type of animal, the therapsids. Not quite a mammal, they had nevertheless developed characteristics that have been passed down to us today: a secondary palate that conveniently enabled breathing and chewing at the same time; nasal

38

structures that allowed the intake of more oxygen; and a posture that allowed for more flexibility. Therapsids begat the cynodonts: small, burrowing, furred animals that are recognisably mammalian. Clutches of fossilised juvenile cynodonts show them curled in an embrace with an adult, suggesting that—even at this early time— they probably exhibited parental care.[14]

Earth teemed with life. But then came a devastation so great that it very nearly annihilated all of it.

———

Extinctions seem to be a hallmark of life on Earth. There have been at least 17 extinction events[15] since 550 million years ago when life exploded in diversity. That's about one every 30 million years or so, though they are not that evenly organised, and only five have really threatened life on Earth—six if you count the current Anthropocene extinction. The greatest of all extinction events occurred at the end of the Permian period, 252 million years ago. This event was so devastating that it closed not only the geological period of the Permian but also the entire geological era of 'old' life, the Palaeozoic, which had lasted for 300 million years. Ninety per cent of life in the sea and 70 per cent of life on land was obliterated, along with the *Glossopteris* forests of Australia, India, eastern Antarctica, southern Africa and South America that had dominated half of the world. It is the only recorded mass extinction of insects, with eight of the 27 orders of insects disappearing.[16]

The end-of-Permian extinction event was actually a series of events: about 19 of them, separated by millions of years, and peaking at around 252 million years ago. Some of the episodes lasted 200,000 years, and some were on hundred-year timescales matching the Anthropocene.[17] The rough outline of events is that a colossal plume of magma—something the size of a planetary moon, say a small moon of Jupiter—rose from the centre of Earth beneath Siberia in a place now known as the Siberian Traps.[18] It spread out under Earth's crust over an area of around 800 kilometres, emplacing about 8 million cubic kilometres of oozing magma, melting the crust and belching deadly bursts of gas.

In an example of monumental scientific understatement, palaeoclimatologists call these events 'excursions'. During an excursion, billions of tonnes of gases such as carbon dioxide were blasted out. Traumatised plants were washed into lakes while they were still breathing. The atmospheric signature locked into their wheezing stomata, like a message in a bottle, reveals an atmosphere pulsing from carbon dioxide levels of 400 parts per million (similar to today's levels) to a staggering 8000 parts per million—levels not seen since multicellular life evolved[19]—and back again over periods as short at 130,000 years.

To make things worse, oxygen in the atmosphere was locked away in a molecular embrace with methane and hydrogen sulphide. Even at sea level, animals suffered mountain-sickness maladies such as pulmonary oedema caused by low oxygen levels.[20] High-altitude regions were probably uninhabitable. Large expanses of ground water, sea water and fresh water became anoxic—deadly to most organisms except for oxygen-loathing anaerobic microbes.

Hard as it is to imagine a world like this, windows to it can be seen in today's 400 human-made dead zones.[21] Think of the Baltic Sea, Chesapeake Bay, Lake Erie and the Gulf of Mexico.

Ozone destruction allowed the land to be irradiated. Seas flooded over the land and drowned habitats. Meandering streams erupted into gouging rivers that clawed at the landscape. Toxic trace metals such as lead and nickel dissolved into, and poisoned, water. The sky rained acid, and acidification killed ocean life. In addition, oceans spiked to a mind-boggling 45 degrees Celsius,[22] six degrees hotter than the recommended temperature for a modern hot tub. A relaxing soak in this sea would have induced heat stroke.

This runaway roller-coaster of kill mechanisms saw each planetary belch become its own extinction event. In southern Africa, at least three pulses of extinctions were recorded during one of the excursions.[23] First to go was a genus of amphibians, the *Uranocentrodon*, which succumbed to a period of extreme arid-ification in the rapidly heated world where ponds and lakes dried up, leaving glaring evaporitic gypsum. Second came the loss of the *Glossopteris* forests. Herds of herbivorous dicynodonts collapsed into a population bottleneck before limping towards a slow recovery. The large carnivorous gorgonopsids that preyed on them—the apex predators of Permian ecosystems, which reached the size of a modern rhinoceros—were completely annihilated.

Life spluttered, winking on and off in the prolonged period of toxic conditions. It took millions of years to recover: the fossil record shows a ten-million-year coral gap; on land, with no forests to make peat, there was a synchronous coal gap. Bravely, life would make valiant attempts at recovery, particularly in the ocean where ammonoids (think of nautiluses) and conodonts (think of eels) evolved rapidly in between volcanic pulses—only to be decimated again and again by further blasts of concentrated carbon dioxide, global warming, acid rain and acidified oceans.[24]

In Australia, rocks of the Sydney Basin show a pronounced extinction pulse at 252.31 million years ago. This pulse marks the complete extermination of *Glossopteris* forests. It must have been eerie. According to the rocks, there was no massive upheaval to wipe life from the surface of Earth. If anything, the climate became a little balmier for a short time. Nevertheless, the forests disintegrated. And quickly. A telltale spike in fungus[25] speaks of withered and sick plants, slimy with algae, in a decomposing muck of dead forests. High nickel concentrations, found in the rocks, appear to have been the culprit. While nickel is an essential micronutrient, in high concentrations (such as might be found in volcanic outgassing) it is toxic to plants, interrupting photosynthesis and respiration.[26] Exposed and unprotected, *Glossopteris* leaves withered and dropped lifeless to the ground.

—

There were survivors: tough horsetails in the marshes, club mosses, ferns and individual conifers. Many of these plants were short-lived communities—disaster flora—that invaded from elsewhere or evolved rapidly to fill empty niches. They were temporary replacements that overprinted the former stable regional flora. A weedy seed fern, *Lepidopteris callipteroides*, left its mark as a fossil in rocks of this time. A splay-legged, rotund, tusked and turtle-beaked synapsid herbivore called *Lystrosaurus* was found in plague numbers; in places they comprised nearly the total mass of fauna grazing on the weedy ferns and mosses, and on the mats of algae that covered putrid ponds. Looking like a dwarf hippo, except for the beaked face, this dog-sized synapsid probably survived in burrows where it was able to fall into an energy-saving torpor. Fossils of the time from the Bowen Basin, 500 kilometres south of Thiaki,[27] show the area reduced to a slimy mire of weedy plants grazed by desultory *Lystrosaurus* survivors and occasional amphibians presumably nosing through algae-covered bogs.

Weedy ecosystems thrived for a while, but they could not form the basis of a balanced ecosystem. Secreted in burrows, caves or holes, tucked in refuges or hidden in pockets against the flanks of mountains or deep in valleys, some bedraggled refugees did cling to life. It was these who laid the foundation for our modern-day fauna of reptiles, mammals and birds.

It would take time—in eastern Australia, it would be another ten million years before the forests returned—but eventually lush forests, with brand new plants, did return. The seed fern *Dicroidium* became the new signature plant, replacing *Glossopteris*. *Dicroidium* had a forked stem, a thick wax covering on its fronds and a complex reproductive cone. Alongside *Dicroidium* was a range of new seed plants and conifers: cycad-like Bennettitales, and Gnetales, gymnosperms that had developed a new system of conduits to transport water—a system that later flowering plants would adopt. The herbaceous understorey once again became rich in ferns and lycophytes, and plants became so well adapted to particular soil types that it is now possible to tell the fossil soil types by the plants that grew there.[28] A new suite of animals—archosaurs, lizards, crocodiles, dinosaurs and birds, cynodonts, and the ancestors of modern frogs and turtles—had also evolved from survivors of the extinguished Palaeozoic.[29]

Fossils in the cliffs around Sydney capture a time ten million years after the end-Permian killing time. These fossils are not only of the bones of animals but also their coprolites (poo). Scientists, as they do, classified these antediluvian droppings into shape categories: some were cylindrical, that is, elongate and tapering at the end, curved or straight, 96 by 25 millimetres; some were ovoid, though with flattened or rounded extremities, with maximum dimensions of 110 by 778 millimetres; some were triangular, 55 by 33 millimetres, with irregular surfaces due to undigested food; and some were quite small and spiral with about nine turns. All in all, there were 11 distinct stool morphotypes, depending on how the shape narrowed or tapered when exiting the sphincter. This painstaking palaeontological proctology reveals that on the south-eastern peninsular extremity of Pangaea, where Sydney is today, close to the austral polar circle, lived a rich vertebrate assemblage including amphibian temnospondyls (ovoid droppings); apex predators, proto-reptiles or proto-mammals (long, tapering droppings); and fishes (spiral droppings).[30]

The populous synapsid dicynodont, *Lystrosaurus*, which had lived on the ground, climbed the trees and roamed the plains in vast herds—and was one of the greatest survivors of all time—dwindled to finally become extinct during the following Triassic.

Being the dominant animal is no guarantee of survival during an extinction event.[31]

———

Forest collapse, ocean acidification, continent-wide stratosphere-puncturing super storms, heatwaves and massive wildfires are all hallmarks of planetary destruction. It sounds eerily familiar and, soberingly, the greatest of Earth's extinction events has its best analogue in today's Anthropocene.[32] Our emissions of astronomical amounts of carbon dioxide are already amplifying into feedbacks caused by global warming. The carbon dioxide blasted out of the Siberian Traps over one 200,000-year period is estimated to be 170,000 gigatonnes.[33] Given the current rate of carbon dioxide emissions, this will take us a mere 5000 years.

This is terrifying territory. Today, a million species are threatened with extinction, and in a world that is predicted to be 4 degrees Celsius warmer as early as the 2060s,[34] between 40 and 70 per cent of the world's species will be annihilated. One-quarter of Earth is already completely degraded, and three-quarters have been altered by humans. Once again, life on Earth is being pushed to the edge of extinction.[35]

Thiaki had a close call recently. Fires swept through nearly a quarter of Australia's forests,[36] pumping an estimated 350 million tonnes of carbon dioxide into the atmosphere—one per cent of the total global carbon emissions of 2019. During a record-breaking heatwave in the Wet Tropics, a rainforest area 37 kilometres south-east of Thiaki caught fire—and no one noticed for nearly two weeks. It was thought that wet rainforests were immune to fire, but we soon learned that it was possible for them to burn.[37]

Further south across the border into New South Wales, the rainforests of Terania Creek—part of the World Heritage-listed Gondwana Rainforests of Australia—burned for the first time in over a thousand years; 1300-year-old trees fell to the ground sheathed in flame.[38] Half of the remnant old-growth forest was burned, killing hundreds of thousands of ancient trees that had been finely ringed with memory.

# CHAPTER FOUR

# A million years of rain

While I was collecting in the Cairns district in North Queensland, during the summer of 1904–5, I was told of the occurrence there at rare intervals of a dragonfly of such enormous proportions that I scarcely credited the story. It was said to come swooping down 'like a bird', and local residents went so far as to declare that 'its bite would pretty well kill you'.[1]

R.J. Tillyard

Geoff Monteith walked back to the Thiaki homestead, slogging up a steep slope with a butterfly net in one hand. His wiry frame bowed under the weight of a heavy backpack stuffed with specimens and gear for observing and trapping insects. Suddenly, a large and striking dragonfly flew tauntingly close to him. With amazing agility and speed, the now retired leading Australian entomologist lunged and snatched it with his hand from the air, his free arm as whip-quick as a snake strike. 'It flew straight in front of me,' he said, shaking his head with amazement. With a wide grin, he showed me an extraordinary specimen of the world's largest and heaviest dragonfly, *Petalura ingentissima* (giant petaltail), which had bitten him several times, drawing blood.[2]

This is not a common dragonfly, so for it to have flown so flagrantly, as if thrown like a dart, within reach of one of Australia's pre-eminent entomologists was, I thought, just one more of Thiaki's little taunts.

With a wingspan of 16 centimetres and body length of about 12 centimetres, the dragonfly was an impressive creature: black with contrasting yellow markings on the head and thorax, plus double yellow bands defining each segment of the abdomen, which ended in the distinctive petal-shaped anal appendage that gave the creature its common name of giant petaltail. Each clear panel of the transparent wings was outlined with black nerves, and its strong legs were large, long and jet black. The dragonfly's stocky larva lurks like a predatory spider in a burrow drilled out of stream banks, venturing out at night to prowl for insects.[3]

Geoff was at Thiaki to advise a small team of researchers who were using the presence of dung beetles to assess the biodiversity of our restoration effort.[4] These little shit-eaters happen to be brilliant bioindicators because they mediate functions that are important to the forest: they decompose dung, help with seed dispersal and survival, distribute nutrients around the forest ecosystem, and improve aeration and water penetration. Geoff could look at a collection of dung beetles—which are fussy about where they live—and tell exactly what environment they came from: rainforest, paddock, new restoration sites or old. This selectiveness would reveal how our restoration efforts were tracking. Were rainforest dung beetles moving into the reforested paddock? Geoff and the dung-beetle team systematically set up little pitfall traps in the intact rainforest and in the restoration plantings, using baits of agile wallaby dung or rotting mushrooms. Apparently, these baits are wildly alluring to dung beetles, which would scurry to secure the treats only to fall into a tube of preserving propylene glycol and water mixed with detergent.

Beetles claim almost every niche on land. They are Earth's most successful animal, with around 400,000 species—a little less than half of known insects. Beetles, it seems, got ahead of the game by marching in lockstep with the first forests[5] and quickly developing the ability to fly, so that they could move readily from plant to plant.[6] They also took advantage of the piles of dinosaur dung, which may have spurred dung-beetle evolution over 120 million years ago.[7]

The giant dragonfly was an incidental bonus for Geoff.

———

The quintessential image of life in the early forests of Earth is one of gigantic dragonflies—three times the size of Thiaki's comparatively petite petaltail—droning through boggy fern forests. Geoff would have needed a lasso to bring one down. But while they thrived 300 million years ago, their fossil trail had gone cold by the end-Permian extinction. It remained that way for the entirety of the following Triassic, from 250 to 200 million years ago.

Genetic analysis of dragonflies,[8] however, reveals a more subtle picture. There had, in fact, been a massive radiation of dragonflies at the beginning of the Triassic that filled the niches emptied during the extinction event. But these dragonflies were small. Triassic creatures began life in a Lilliputian world where only the small and fleet survived by rallying through the smoke and toxic haze to forge the foundations of the Mesozoic—the new era of 'middle life' that arose from the scraps of old life from the Palaeozoic. As with other extinction events, the empty spaces were initially flooded by a weedy disaster flora and fauna of fungus, ferns and *Lystrosaurus*. But struggling alongside the bedraggled dragonflies were some amphibian temnospondyls, some tiny primitive lizards and the first of the reptilian archosaurs.[9]

The pattern of life had once again lurched onto a different track.[10] The reptiles—which had been widespread, but skulking in the background and biding their time—now took over the world. They prised loose the hold of the weedy disaster fauna to increase in extraordinary number and diversity: there were long-snouted semi-aquatic phytosaurs, heavily armoured aetosaurs, sleek and predatory reptilian ornithosuchids, and carnivorous pseudosuchians (just imagine a large crocodile, but rather than walking with its legs splayed to the side, it was stomping about on hoofed pillar-like legs). Gone was *Lystrosaurus*, and in its place were the new rhynchosaurs: pig-sized hammer-headed reptiles.

The flora also had a complete shakedown. Gone were the *Dicroidium*-dominated forests. By the mid-Triassic, 220 million years ago, Thiaki was a richly layered Gondwanan forest of closely spaced podocarp conifers with spirally arranged needle leaves, like the still-living *Dacrydium cupressinum* (rimu), a New Zealand conifer that grows up to 50 metres tall. There were also *Araucaria* conifers, tree-sized horsetails and smaller heath-like stands of regionally endemic seed ferns, as well as a dense ground cover of ferns such as *Cladophlebis*.[11] This was so widespread that it might have been a Mesozoic weed. There were also Bennettitales (seed plants that looked a bit like cycads but had flower-like reproductive parts), reeds in swamps, and other scrubby cycad-like plants.[12] The flora was remarkably similar across Pangaea, suggesting an overall equable climate.

Moving stealthily through the Triassic foliage, however, was also something quite new: diminutive and fleet-footed, with erect limbs under the body that created an upright posture, they were the dinosaurs.

———

In 1983, footprints in a paper-thin layer of dark grey shale in the spoil heaps of an open-cut coalmine[13] in south-east Queensland surprised students searching for plant fossils in the 230-million-year-old layers of the Ipswich Coal Measures of a middle Triassic age called the Carnian. The unexpected three-toed footprints were made by a new kid on the block: an early carnivorous dinosaur, possibly *Eoraptor*. With an elongated skull housing an arsenal of serrated and recurved teeth, and the feet and 'hands' wielding sharp claws, *Eoraptor* was not only one of the earliest Gondwanan dinosaurs[14] but also one of the earliest ancestors of birds. Active predators— light and graceful—they scampered about, avoiding notice in a world dominated by the speciose archosaurs.

But then, at around 232 million years ago, it started raining— and it didn't stop for over a million years. The period, known as the Carnian Pluvial Episode,[15] is captured across the Pangaean world as a thin band of grey rock running through the heart of the desert- red stone that is normally the hallmark of parched landscapes. The rocks speak of a world that had jolted quite suddenly from arid to swampy[16] as a result of an enormous outpouring of lava, which piled up to 6 kilometres deep in less than two million years in the vicinity of north-west North America.[17]

The resulting greenhouse gas emissions warmed the desert world by 7 degrees Celsius. Hot continents drew in oceanic moisture, which whipped already strong monsoonal rainfall into mega- monsoons that ate up deserts and flushed dry vegetation-free valleys into braided rivers carrying sandy and pebble sediments. When geologists looked, they found evidence of a wet episode everywhere, in layers of sand and conglomerate, or in offshore reefs smothered by sediment that caused the extinction of many corals.[18]

Changing climates cause extinction. And, sure enough, when it stopped raining and the world turned arid once again, it was a different place. While not on the scale of one of the big five extinc- tion events, the Carnian Pluvial Episode nevertheless caused a major reshuffling of life: many of the ruling archosaurian reptiles took a beating. The dominant rhynchosaurs, which had formed 50–90 per cent of fossil specimens in Triassic sediments, dwindled. But dino- saurs, those small furtive creatures that evolved in the turmoil of recovering life after the end-Permian extinction, and entered the

Carnian as underlings, soon emerged the winners: from being virtually invisible, they overwhelmingly dominate the fossil record of the time.[19]

———

Two main groups of dinosaurs became dominant: the ornithischians (which would evolve into dinosaurs such as *Stegosaurus* and *Triceratops*), and the saurischians (which would evolve into dinosaurs such as *Brachiosaurus*). The saurischians included the theropods, which would lead, much later, to *Velociraptor* and *Tyrannosaurus rex*. For 30 million years, leading up to the end of the Triassic, dinosaurs thrived and multiplied, as did some of their non-dinosaur reptilian competitors who might have maintained the upper hand on Earth today if not for another volcanic outburst: this time the labour pains of the incipient Atlantic Ocean.

In one of the greatest flows since the creation of complex life on Earth, molten magma pushed against 10 million square kilometres of Earth's crust—across Africa, South America, North America and Eurasia—and sent lava oozing across Pangaea. One of the big five extinction events, the lava flows of 200 million years ago closed the Triassic and ushered in the Jurassic with an afterbirth of massive amounts of carbon dioxide and ocean acidification.

While many archosaurs became extinct, some dinosaurs managed to limp through the 600,000 years of upheaval. For whatever reason—differences in locomotion, growth or metabolism—their lifestyle gave them the edge.[20] As in other extinction events, the survivors were small and swift. In this case, it was the bipedal dinosaurs: the petite predatory theropods that would spawn *Tyrannosaurus rex*, and ultimately birds; the dog-sized ancestors of sauropods that would later drop onto all fours to become the largest dinosaurs to walk the earth; and the bipedal ornithischians that would evolve into the horn-faced dinosaurs such as *Triceratops*. A tiny leaping archosaur, looking like a cross between a lizard and a frog, would evolve into the flying reptilian pterosaurs,[21] with wingspans as wide as a small plane. All other archosaur lineages—phytosaurs, aetosaurs, ornithosuchids, and pseudosuchians—vanished.

Dinosaurs didn't waste any time in their quest to become big. On Antarctica's Beardmore Glacier, the same glacier where Scott found *Glossopteris*, a younger rock contained an early Jurassic fossil from the largest-ever theropods. *Cryolophosaurus ellioti*[22] was about 7 metres long and weighed about 465 kilograms. It lived a mere ten million years[23] after the extinction event, having, in that brief time, probably evolved from a small dinosaur.

Pangaea started to disintegrate during the Jurassic: it began to gyrate, with different continents wheeling at different speeds and then in different directions, causing the great landmass to tear along huge rift valleys and spill lava like blood. The Panthalassa Ocean Plate—the future Pacific Plate—continued to plunge at an angle beneath eastern Australia, raising the east coast. Australia rotated away from the South Pole and into slightly lower latitudes. Thiaki basked in a warmer climate. Dinosaurs roamed the land, and pterosaurs soared in the Jurassic sky.

Blemishes on fossilised plants show that the time was literally buzzing: the plants were nibbled by already evolved beetles and grasshoppers or katydids; pierced by bugs, flies and wasps; and the litter was munched by cockroaches.[24] Frogs, feasting on the abundant insects, had long before dragged themselves through the end-Permian killing time, which reduced 24 proto-amphibian families to eight.[25] But by the Jurassic it was a noisy place, perhaps echoing the sounds of today—except for the roaring, grunting and piping of dinosaurs.

Placid lakes and swamps—surrounded by a rich coniferous forest with a dense heathy ground cover of ferns, cycads and seed ferns—allowed several species of large carnivorous theropods from emu-sized up to the size of *Tyrannosaurus rex* (who was still 90 million years in the future) to come of age.[26] They stalked the herds of grazing archosaurs and newly evolved herbivorous dinosaurs. In the freshwater lakes that lay to the south-west of Thiaki were plesiosaurs, including rare freshwater species that inhabited the fresh to brackish basins to the west. Having survived the end-Triassic extinction, plesiosaurs quickly commandeered the emptied seas and bays of Gondwana, eventually finding their way to an embayment south-west of Thiaki and evolving into the oldest freshwater species of plesiosaur in the world.[27]

Life, as it does, was once again filling Earth's ample niches. Hidden from sight of the rambunctious dinosaurs, in the understorey and in burrows, were the early mammals. These had evolved from cynodonts, a proto-mammal that had managed to scrape through the end-Permian extinction event, cowering underground, in the form of a small dog-like burrower. Fifty-seven genera of cynodonts have been found with a 'Gondwanopolitan' distribution in South America, Africa and Antarctica,[28] presumably ambling across to Australia, although no fossils have been found yet. Cynodonts flourished in the Triassic, growing to different sizes and evolving into both carnivores and herbivores. But the drenching, waterlogging Carnian rain was their undoing.

While dinosaurs diversified—there are 547 known genera to date,[29] including at least 1000 species[30]—the mammals oozed into niches not occupied by reptiles. They accumulated innovations that led to a short-fuse burst of evolution, branching into 20–40 major groups of an evolutionary bush that is difficult to tease out, including over 4000 fossil genera. Today, there are 5400 species of mammals. In their explorations of empty ecospace and novel adaptive zones, many mammals became nocturnal, burrowed, swam or climbed trees (from where some would launch into the air as gliders).[31]

Within Australia's Jurassic parklands, Thiaki's forests were rich with as many as 200 species of plants[32] including araucarian species similar to today's *Agathis microstachya* (bull kauri). Abundant newly evolved seed plants, displaying innovative reproductive organs, grew on the slopes. Tree ferns, ferns and lycopods created a dense understorey. Forests grew densely in swamps and basins, which were slowly smothered by sediment from the rising eastern highlands. Deeply buried, these massive carbon-based graveyards (such as the Surat Basin, part of the Great Artesian Basin) are today exhumed for coal and gas.

———

A faded black-and-white photograph taken in the early 1960s shows a child with long skinny legs, a pudding belly and dark brown hair, severely cut. I was an outdoors child, a tomboy. In a decade of summers, some memories stand out clearly: a burrow in a nearby field, where I found a nest of baby mice with their eyes still shut (I remember tearing through the straw-yellow grass in rubber-thonged feet, bursting to display the limp creatures to my stoic mother); glass jars full of tadpoles that, if they weren't extinguished in a careless anaerobic morass, transformed miraculously into frogs; snails whose pedicelled eyes obligingly retracted and extended each time they were poked; and the seasonal blossoms of almond trees, which I picked daily despite the fact that they would bruise and fall before I got them home. I could never sit still to read a book. Inanimate dolls were lost on me: I was given one once; it sat on the top of the cupboard, peering out through blue plastic eyes. Better to be outside where the wind whispered among the leaves and sent waves undulating through fields of grass.

One day, I was so taken aback by the beauty of a thing that I simply couldn't touch, poke or pluck it. A teacher on a field excursion had shown me a grevillea flower with the bewitching name 'lady-finger'; it was an immaculate, silky, rosy pink flower with a complicated shell-shaped floral tube split into perfect lobes curling like chaos, out of which emerged a slender curved style, the 'finger'. I stared and stared at it, trying to comprehend how such a thing, so evanescent, so complicated, so perfect, could come to be. Similarly, the delicate imprint of a fossilised plant on a rock held such fascination that, after learning that fossils were found in quarries,

I meticulously examined the quarried gravel in our driveway for weeks looking for imprints of life. The tiny remnant patch of scrub on the top of the nearby hill was my special place: one of dappled sunlight, twisted tree limbs and, in season, beautiful wildflowers. It was to be approached only from a certain direction, skirting around the expanding patch of unfriendly weed that grew darkly and spread menacingly outward. No birds alighted on or near the weed patch, no lizards scurried in between, strangely nothing else except for grass would grow in the patch, and even the grass was cautious.

At the local agricultural show, glorious bunches of wildflowers were on display, hoping to catch the eye of the judge and be awarded the cherished scrap of synthetic blue ribbon of first prize for wildflower arrangement. When I was a small child, I could have picked such a bunch on the scrubby hillside. They had disappeared completely by the time I was a teenager—a tiny extinction entirely enabled by me.

———

> The rapid development as far as we can judge of all the higher plants within recent geological times is an abominable mystery . . . I should like to see this whole problem solved. I have fancied that perhaps there was during long ages a small isolated continent in the S. Hemisphere which served as the birthplace of the higher plants—but this is a wretchedly poor conjecture.
>
> Charles Darwin, 22 July 1879[33]

Darwin, as usual, was correct—it was a poor conjecture.

Terrestrial plants, though, had hit the ground running. By the time fishes heaved themselves out of water on lobed fins to become the first sluggish amphibians, about 370 million years ago, at the end of the Devonian, plant innovation was already mostly done and dusted.

From an almost-nothing algal film of green sprang plants with water-conducting cells, cuticles to protect surfaces, stomata to breathe, sporangia to produce spores, and rhizoids to bother the earth. With this basic suite of adaptations, plants grew in complexity and size: water-conducting cells locked into vascular columns; organs such as leaves and roots emerged from exploratory outgrowths and rhizoids; vessels carrying water thickened into wood, bracing the trees that soared skyward; and sporangia evolved into seeds.[34] Forests of trees up to 20 metres high covered Earth.[35] There was only one thing missing: flowers.

They burst onto the scene in the fossil record almost miraculously—as if by the hand of God. It was too much for Darwin, who was in the process of murdering God with the idea of evolution. All hell broke loose when he published the theory of evolution with Alfred Russel Wallace in 1858, especially after his publication of *On the Origin of Species* the following year, which Pope Pius IX condemned by placing on the Index Expurgatorius.[36]

With so much skin in the game, it was imperative that the flower problem be resolved. Darwin grumbled that fossils of flowers did indeed seem to appear in the fossil record almost instantly and almost everywhere, at least in the northern hemisphere. His last words on the matter in 1881 were that he 'fancied that the development might have slowly gone on for an immense period'.[37]

To this day, flower evolution remains an intransigent problem. Flowering plants do, in fact, appear in the fossil record suddenly— and quite recently as far as evolution goes, in the early Cretaceous, around 135 million years ago. But molecular evidence and new ways of modelling fossil ages[38] strongly suggest a long-fuse evolutionary history sparked much earlier. It is even possible that the specific seed plant that would spawn angiosperms had emerged as far back as the Devonian, when Thiaki still lay under the sea.

In terms of the tree of life, flowering plants must surely be both the trunk and the crowning glory. They occupy almost every habitat on Earth and choreographed the rest of today's biodiversity: their evolution set the tempo of much of life on Earth. When angiosperms appeared on the scene, they rallied insects to their cause like politicians with enticements and bribes. The majority of the 352,000 species of angiosperms are pollinated by animals,[39] mostly by insects. And about a quarter of all insects feed on flowering plants, including half of all beetle species, and nearly all of the hundreds of thousands of species of butterflies and moths; all bee groups rely on flowering plants. Bees and angiosperms diversified at similar times. Indeed, over 22,000 flowering plants have tube-like anthers that carry pollen, which is released when a female bee bites into the anther and then buzzes, or sonicates. About 58 per cent of all bee species sonicate flowers to extract pollen.[40]

The relationship between plants and insects is not something new—300-million-year-old fossil insects have been found with masses of pollen in their guts—and occurred long before the advent of flowers. Cycads, for instance, need weevils (a type of beetle) for pollination, presumably an arrangement that was negotiated with beetle ancestors perhaps as long as 300 million years ago. A specific genus of moth, *Agathiphaga*, feeds only on the seeds of Thiaki's coniferous *Agathis microstachya* (bull kauri), a relationship that has been 200 million years in the making.

*Eupomatia laurina* (bolwarra), which grows along the east coast of Australia, including in Thiaki, is a species from the archaic family Eupomatiaceae. Evolving before modern birds, *E. laurina* lacks sepals and petals, instead having leafy bracts protecting the ovule. The outside of the flower is fringed by sprays of spirally arranged stamens, and petal-shaped staminodes cluster in the centre. With a strong, fruity-musky fragrance, these lure a specific type of weevil. First thing in the morning, as many as 80 weevils gather within the small, 3-centimetre flower. They spend their day mating and munching on the soft tips of the staminodes, which also provide a protective yet airy roof. In the late afternoon, the weevils leave the flowers from between the sticky staminodes, carrying heavy loads of pollen to other flowers. The weevils lay eggs in the flowers, and when the stamens and staminodes fall to the ground, they take with them the weevil larvae. These feed, away from the growing seeds, on the fallen remains, afterwards pupating in the soil.[41]

———

A few years ago, a brief botanical survey of Thiaki's trees produced a list of 220 species,[42] virtually all of them angiosperms. This is a little less than 20 per cent of the tree species of the Wet Tropics. (Thiaki's herbs, shrubs, vines and epiphytes were not surveyed—in the Wet Tropics as a whole, these nearly triple the number of vascular plant species,[43] which amounts to about 3000 species.) Thiaki's diversity is not particularly special on a world scale, but this diversity is as close to as many species as this little patch has ever had.

What is extraordinary, however, is the presence of so many ancient plants. In fact, about half of the surveyed list are old or endemic species. Antiquity was a key attribute that allowed the inscription of the Wet Tropics of Queensland onto the World Heritage List in 1988. Thiaki's subset of plants mirrors the world-heritage quality of the Wet Tropics, containing

> one of the most complete and diverse living records of the major stages in the evolution of land plants, from the very first pteridophytes [vascular spore-bearing plants like ferns] more than 200 million years ago to the evolution of seed-producing plants including the cone-bearing cycads and southern conifers (gymnosperms), followed by the flowering plants (angiosperms). As the Wet Tropics is the largest part of the entire Australasian region where rainforests have persisted continuously since Gondwanan times, its living flora, with the highest concentration of primitive,

archaic and relict taxa known, is the closest modern-day counterpart for Gondwanan forests.[44]

As it turns out, this description is misleading because it implies that flowering plants evolved from gymnosperms. The fact is that no one actually knows where flowers came from—still. Angiosperms have vessels carrying water and nutrients, plus broad leaves with a reticulated pattern of veins. However, the protection of ovules is the definition of angiospermy. In flowering plants, ovules are protected in the carpel of the flower. Pollen germinates on the style, and a pollen tube grows from the pollen down the style to the ovule, where the sperm is released. This controls which pollen sperm fertilises the egg. Bisexual reproductive structures occur only in angiosperm flowers.[45]

In gymnosperms, on the other hand, the seed is naked—the word gymnosperm literally means 'naked seeds'. The entire reproductive cycle in gymnosperms is comparatively slow: it may take a year between pollination and fertilisation, and another year for seeds to mature. In flowering plants, fertilisation may happen in hours and seed maturation in days or weeks. The tendency to protect the ovule, however, has been a part of plants for a long time. Even the glossopterids that cloaked most of Gondwana and were the dominant forest of Thiaki before the end-Permian extinction protected their ovules within a lamina-like structure.

The current accepted theory of flower evolution is that both gymnosperms and angiosperms have a common ancestor—which means they both evolved from even more ancient plants. But which ones? For flowering plants (and perhaps for much else), the present is an imperfect reflection of the past,[46] so their ancestors may have looked nothing like what we call flowering plants. In the Jurassic parklands, and perhaps even earlier into the Triassic,[47] it seems that angiosperms were a minor element of the forests, growing in moist disturbed habitats like weeds. They may have had corrugated leaves, which furled into fantastical structures that encapsulated the ovule and were surrounded by towers of fronds extruding fingers of structures that contained pollen. Or they may have had surreal stems that erupted into displays of cupped ovules above other stems of pollen-producing structures.[48]

Thiaki, it turns out, knows what early fully formed flowers looked like. Sitting at the very base of the angiosperm family tree is one of the most ancient families of flowering plants on Earth, Austrobaileyaceae. Only a single species of this family, a liana named *Austrobaileya scandens*, has made it through at least 120 million years to climb and loop along the edge of our back deck, which tucks carefully into the rainforest. *A. scandens* is an extraordinary being. Its seedlings produce two types of shoots. One type meanders about to explore spaces in the shaded understorey. These lively little rods can miraculously re-root when dragged to the ground during a treefall. New shoots can either poke upwards to begin exploration or nose underground for a while before developing a new cluster of shoots. The other type of shoot grows from the seedling in the form of slender tendrils that twine up nearby woody trees towards sunlight. At the right level, the tendrils form a sparsely leaved canopy of cylindrical photosynthetic vines that hang in tresses from trees and harvest the diffuse light of the rainforest that radiates in all directions.[49]

The leathery blue-tinged leaves of *A. scandens* are arranged so that they will not cause any self-shading as the vine searches deep within the twilight forest. Chloroplasts in leaves house factories that turn light into energy for the plant—in *A. scandens*, these factories are arranged in a pattern that preferentially absorbs the energy-efficient red-green light of the dim rainforest floor, reflecting back the inefficient blue light.[50] Leaves burn in direct sunlight and thrive instead in the dappled sun of the deep forest. Here, cellular structures in the chloroplast employ quantum mechanics to ensure that life is squeezed out of every last photon by slowing down the speed of light.

The 5-centimetre wide flower is greenish yellow with brown spots, and it hangs from the leaf axis. It beautifully resembles a science-based computer-modelled ancestral flower:[51] it is bisexual (having both male and female parts) and composed of tiles of spiral bracts, sepals and petals, and numerous stamens and staminodes. The flowers emit a rotting odour not unlike the fishy farts of our cattle dog, whom I have on occasion unjustly accused of the smells emitted when *A. scandens* is flowering. Evolving before bees and butterflies (and cattle dogs), the plant has ensured that this putrid smell is attractive to flies, which pollinate the flowers.

You won't find *A. scandens* on the edges of roads, clearings or paddocks. The vine is one of the forest twilight, suggesting that angiosperms started out as plants that exploited dark and wet places—as well as disturbed habitats, thanks to dinosaurs.[52] For more than half of its evolutionary lifetime, *A. scandens* lived under constant threat of trampling, gouging and being ripped in half by dinosaurs, the dominant vertebrate herbivores. Dinosaurs had fantastical solutions to eating vegetation: some chewed or sliced their food; some had beaks with self-sharpening cheek teeth to pry open tough cones; others had gastric mills—strong muscles concertinaed into food-absorbing ridges—or long digestive tracts and slow passage times to break down even the toughest of plants.

Most of the herbivorous dinosaurs were large. With body masses in excess of a tonne, they needed 200 kilograms of plant matter a day. No wonder *A. scandens* has built-in bayonet joints along its stems that facilitate quick disengagement and ready resprouting from snapped stems. A scientific paper titled 'Did dinosaurs invent flowers?'[53] even posited that herbivorous dinosaurs had a major role in the rise of angiosperms. This, however, doesn't seem to be the case because the evolutionary innovations of dinosaurs and angiosperms don't coincide. It appears that angiosperms simply kept their heads down during the reign of the dinosaurs. With dinosaurs apparently lacking the sensibility to appreciate the miracle of flowers, angiosperms stayed little and discreet with small seeds that could quickly exploit dinosaur-trampled habitats. The dense multistoreyed forests that we know today, with large seeds and fruits that are characteristic of closed forests,[54] emerged only with the demise of the dinosaurs at the end of the Cretaceous 66 million years ago.

———

One day while exploring Thiaki's hidden nooks and crannies, I stumbled across a short yet steep valley on one of the far edges of the rainforest. Within a few steps I'd gone from the glare of the grassy yellow paddock into dappled green light, the forest floor a palette of different colours, each fallen leaf a new brushstroke; the soil was a living ganglion of humus. Walking along the ridge edging the valley, I suddenly came upon a stupendous tree taking refuge deep in the fold. A singular presence, its buttress roots flung across the valley and claimed the space as its own. Clambering down to the tree, I found that the forest floor was strewn with fruits that had been gnawed by giant white-tailed rats. There were scratches on the tree's mottled bark made by possums and tree-kangaroos. One of the tree's flying buttresses enfolded a pool of still water that was cool and clear. It was a near-magical oasis.

I pressed my back against the tree and let my thoughts drift. I recalled a poem by Bill Neidjie, a remarkable Aboriginal elder, who mediated Aboriginal knowledge that embraced the singularity of nature and humans.[55]

> Well I'll tell you about this story,
> About story where you feel . . . laying down
>
> Tree, grass, star . . .
> because star and tree working with you.
> We got blood pressure
> but same thing . . . spirit on your body,
> but e working with you.
> Even nice wind e blow . . . having a sleep . . .
> because that spirit e with you

Listen carefully this, you can hear me.
I'm telling you because earth just like mother
and father or brother of you.
That tree same thing.
Your body, my body I suppose,
I'm same as you . . . anyone.
Tree working when you sleeping and dream.[56]

Aboriginal people understood that humans, animals and plants had a common origin through the ancestors, and that ancestors sometimes transformed into plants. Some important trees are countrymen; people had rules for accessing plants, and sets of relationships with certain plants.

Tree and me. For most of the four billion years or so since the beginning of life on Earth, Tree and I shared a common ancestor; we were forged together on a fiery planet and rode the vicissitudes of life through the cooling of Earth from a Hadean molten state resembling Venus to states of frigid glaciation. Our atmosphere in the early days was poor in oxygen and rich in hydrogen sulphide and carbon dioxide. Together, we witnessed the capture by bacteria of light energy, which was spun into chemical energy to drive the transformation of carbon dioxide and hydrogen sulphide into energy-rich food. Oxygen was a waste product that was toxic to life on Earth then, and very nearly annihilated it. The oxygen-facilitated ozone layer, however, allowed survivors like us to float near the top of the ocean without being irradiated, and to dream of innovations.

One of our ancestors engulfed another to form a cell with a nucleus, splitting life into those that remained without a nucleus—prokaryotes, such as bacteria—and those with a nucleus—the eukaryotes, such as Tree and me. A relative of our friend the *Rickettsia* bacteria then got in on the act. It began a merger with a eukaryote, offering its ability to metabolise light energy into chemical energy in return for a nutritious eukaryotic soup of metabolites. While the merger didn't start well, with the bacteria said to have parasitised the eukaryote, the end was spectacularly profitable—an organism with a never-ending biological battery now called a mitochondrion, which Tree and I both have.[57]

But then Tree and I parted ways. About 1.5 billion years ago, Tree's ancestor formed a partnership with a green bacterium that offered the technology of photosynthesising sunlight into sugars using water and carbon dioxide. These bacteria cells became chloroplasts, and they live only in plants.

Each of our ancestors gradually became more complex, evolving multicellularity at much the same time so that different cells could take on different tasks. With sunlight for energy, movement was no longer an imperative for Tree's ancestors, so their focus turned towards gathering energy, water and minerals. The design solution for such a lifestyle was repetitive branching structures—organs such as leaves, stems and roots—that could regrow after damage by herbivores. Compared to animals' adaptability, this plasticity was close to immortality—we, for instance, can't just discard the odd liver or heart and plants got a head start on the dominion of land. My ancestors, who were close behind, had lost any ability to synthesise organic compounds from chemicals, thriving instead by eating other life forms (including plants). We needed movement and evolved muscles, sensory organs and a complex nervous system for coordination.

Tree senses humidity, light, minerals, gravity, wind, soil structure and the presence of others—friend or foe—through leaves, stems and roots. Tree is acutely aware of light through photoreceptors related to those of animals; smell through the perception of chemicals; vibrations and sound through membrane proteins. Tree's stomata shrink and leaves shrivel and drop when Tree is cold. Unlike our singular heart, Tree's whole being pumps water, and hormones in the sugar pulse through phloem. The phloem also mediates electrical communication. And a plant's nervous system is based on action potentials just like neurons in animals. These are highly cross-linked to form a complex network: it is entirely possible that these somehow contain memory and facilitate learning.[58]

Tree is smart, making countless daily decisions based on the shifting environmental circumstances to produce the best outcome for Tree's offspring: sprouting masses of roots in newly found resource-rich patches and muscling out competitors. Tree's decisions about branching and flowering are made a year or two in advance, based on a history—a memory—of the environment.[59] I imagine Tree derives some sexual pleasure in pollination, and contentment in swelling fruits.

——

It took me a while to work out Tree's more recent ancestry. The fallen fruits littering the dappled ground were from another species and had been washed down to lodge against Tree's buttresses. The leaves and flowers were always high and unreachable. Finally, I took a botanist friend down to the valley and, with a glance, she said, 'Oh, I know that one.' (Don't you hate that!) 'It's *Karrabina biagiana*, endemic to the Wet Tropics.'[60] *K. biagiana* (northern brush mahogany) belongs to the Cunoniaceae family, one of several families of the overarching order Oxalidales that evolved around 110 million years ago[61] in the middle Cretaceous, when Australia was, amazingly, not a peninsula continent stuck on the edge of Gondwana, but a series of large islands.

By the mid-Cretaceous, around 110 million years ago, Pangaea had disassembled, with India, Africa and South America detaching from Antarctica, each on their own tectonic plates.[62] India was ferried at a clip towards Asia, which lay near the equator, while Africa and South America unzipped and gave birth to the Atlantic Ocean. The slabs of the Panthalassa Ocean Plate, now the Pacific Plate, continued to plunge deep into the magmatic asthenosphere underneath Australia, dragging down the continental crust so that a third of Australia sank. In a warming world with no polar ice and the sea already rising, the ocean spilled into vast basins through openings hundreds of kilometres wide near Brisbane, Cooktown on Cape York, and across the entire Gulf of Carpentaria. Thiaki lay at the northern tip of a long island, about the size of New Zealand, with the southern tip around the border between New South Wales and Queensland. To the north and west were shallow seas, today's Laura Basin and the Great Artesian Basin, and to the east was the Pacific Ocean.[63]

If I had travelled from Cairns to Perth at that time, it would have involved an overland trip across the open coniferous forests of the Great Dividing Range to the ocean lapping its western flanks near Hughenden. The trip would have been best taken in summer to avoid the long cold polar winter. Pushing through dense heathlands of ferns, fern trees, cycads and scrubby angiosperms, I likely would have startled ankylosaurs now called *Minmi paravertebra*. These 3-metre-long armoured dinosaurs romping about on four stout legs were probably endemic to the Queensland island, eating soft ferny vegetation that they nipped off with a beak before slicing and dicing with cheek teeth. Scurrying in the leaf litter would have been small, primitive, rodent-like mammals called multituberculates (named for the multiple nodules on their back teeth), and other mammals the size of a medium dog but which looked more like a bizarre spineless echidna.[64]

Cessna-sized reptilian pterodactyls, such as *Mythunga camara*, would have been circling in the sky on 4-metre-wide wings and occasionally plunging into the sea for bony fish. Other pterodactyls might have been raucously calling from roosts in rocky hills nearby. The shoreline would have been abuzz with insects we'd be familiar with today, and there would have been burrows made by an early form of platypus, *Steropodon galmani*, as well as more modern-looking monotremes. From the water's edge, turtles would have popped up for a look, and crocodiles would have made skid marks as they slipped into the sea.

The sea journey would have been enlivened by the 10-metre-long pliosaur *Kronosaurus queenslandicus*, a carnivore twice the size of a great white shark. And pods of a dolphin-like ichthyo-saur, *Platypterygius longmani*, would have breached nearby.[65] Nosing around the shoreline near Alice Springs would have been amphibian temnospondyls—large enough to stuff a small dinosaur into their mouth—relict populations of a group that dates back to 370 million years ago. The amphibian temnospondyls survived into the Cretaceous[66] only in Australia. Elsewhere, the group had become extinct 200 million years before then. In the evening, lungfish—a primitive fish that breathes air—would have splashed in the shallows with lobed fins, seeking earthworms, small fish and frogs.

I would have noticed that the vegetation of the Central Austra-lian island had different species of canopy conifers and understorey ferns, cycads, tree ferns and seed plants. I would have had to travel overland to the next ocean crossing, launching from a jetty near Warburton, Western Australia. The overland trip would have been a veritable dinosaur safari, with dozens of species of small and large herbivores, some in herds, being pursued by reptilian carnivores. I might have glimpsed rare, small, secretive mammals as they scuttled into burrows or raced towards the top of the open forest canopy.

The sea journey from near Warburton, across the 800-kilometre-wide Gibson Desert sea lane, would have disembarked near Kalgoorlie. Overhead, birds would have drifted. At first glance, they might have looked like seagulls, but closer inspection would have revealed teeth and clawed wings, and a head that resembled a feathered dinosaur.[67] Travel would have been by land to Perth, itself a mere half-day road trip to Antarctica to which Australia remained connected via a spectacular rift valley stretching to breaking point; Australia was biting at the bit to be released into warmer northern climates.

Floating on Australia's cool polar marine ways would have been a fantastic flotilla of Cretaceous flotsam, including migrating flower-ing plants such as the offspring of *Austrobaileya*; they would have come in along the currents from lower latitudes, where flowering plants had already taken hold. India, Africa and South America still lay nearby, having just started their tectonic journey. The Atlantic opening between South America and Africa was a direct channel to Antarctica, with currents ripping along the Antarctic-Australian coastline and into Australia's spectacular seaways. The Tethys Ocean to the north still remained a huge embayment washing the shore-lines of India and Africa to the west, and Eurasia, lying in tropical seas near the equator, to the north. It seems that Tree's family had started as migrants rafting in from Europe.[68] Another thing we have in common.

As Australia's sea lanes drained away, around 100 million years ago, the disturbance-loving angiosperm shrubs and herbs got a root-hold in the swampy and lake-filled landscape. They began infiltrating the scrubby fern-dominated understorey of the open podocarp-araucarian coniferous forests.[69] By 85 million years ago, proteaceous plants had marched across the landmass of Australia from their centre of evolution in Antarctica and southern Austra-lia, and they had become part of the forest canopy. Genera from that time, such as *Macadamia* and *Hicksbeachia*, are still found in today's Wet Tropics. At the end of the Cretaceous, Thiaki was a rich and diverse rainforest with species from families such as Myrtaceae and Sapindaceae; we are planting these in our restora-tion plots.[70]

This fantastic world was about to come to an apocalyptic end as an asteroid headed for Earth.

# CHAPTER FIVE

# Zombie busters

It was as if a thermonuclear bomb had gone off. Pale, shattered, bare limbs were thrown up against a choked sky. Giant trunks were levelled; they lay lifeless on the ground with their roots wrenched from the earth and splayed up from the excavated craters. Rain-gouged gullies were clawed down to muddy streams on steep slopes wiped clean of vegetation. There was not even a whisper of life—the only movement was that of drifting smoke and ash. Weeks later, people spoke of a disaster flora of fungi, ferns and weeds that would plague land management for decades to come.

> There is something deep, and rather frightening, about the history of possession and exploitation of the land in newly settled countries. The thrill enjoyed by some people of possessing and knocking hell out of a virgin landscape is close to the passion of vandalism or rape.[1]

---

Across the Atherton Tablelands came the call for opening up the forest. Pioneer farmers stood in awe, craning their necks to see the canopy of the forest giants against the sky, before bringing them to the ground. The forest was luxuriant, with its trees, lianas and orchids, its myriad birds and strange other-worldly mammals, but the farmers believed that its dark melancholy made the soil sour since no grass grew in this land of shadows and filtered light. Instead, the forests had to be erased to establish order in the landscape and sweeten it with sunlight; the rays would shine on the promised land, with its thousands of closely settled and smiling homesteads. This future was already enshrined in government policy[2] as a yeoman's idyll adopted from an alien England. In the 1880s, the Atherton Tablelands were some of the last accessible rainforests of Australia to fall in the pursuit of such a fantasy.

The clearing frenzy of the first 20 years of the twentieth century brought about the near extinction of the Mabi forest, the main upland rainforest type on the Atherton Tablelands. Nearly all of the Mabi forest was razed.[3] Just scraps—less than four per cent—on rocky ground, which was too difficult to crop, are left. With a higher-altitude vegetation type, Thiaki lay adjacent—scarred by logging, and brooding over the holocaust.

The cedar-getters had already been here first, of course, blazing a trail for the settlers right up and down the coast of Australia. Beginning shortly after the landing of the First Fleet in 1788, convicts were freed and given tickets of leave to make a living as best they could. The roughest of the rough men, wild and desperate ruffians, as muscular as bullocks, they took to the rainforests, living under piled up sheets of bark and moving in search of *Toona ciliata* (red cedar)—or red gold, as they called it. They leaped onto springboards, swinging axes and pulling cross-saws, initially keeping the Aboriginal people at bay with guns and dogs, and then bribing them with tobacco and rum to labour for the cedar-getters. From Ulladulla south of Sydney in the 1820s, working their way north, the cedar-getters erased the red cedar from the landscape; from the

Macleay River, Bellinger River and the Clarence River in 1830s; Richmond in the 1840s; and the Tweed and Moreton Bay in the 1860s. Further into northern Queensland, they hunted the red gold until they reached the end of the cedar rainforests in the Wet Tropics and finally its hinterland, the Atherton Tablelands.

There was more than a faint echo of Joseph Conrad's *Heart of Darkness* about the madness associated with obtaining red cedar. 'You would think they were praying to it. A taint of imbecile rapacity blew through it all, like a whiff from some corpse . . . I've never seen anything so unreal in all my life.'[4]

The wood of the red cedar was soft, lustrous and silky to the touch; rose-coloured when young and deep red when mature, it bordered on black when ancient. The trees were enormous and could be 60 metres high with girths of 11 metres. You could make a lot of money from felling red cedar: £600 per month in the early 1800s. In 1869, a single tree was said to be worth up to £100 in the Sydney markets[5]—in today's dollars, that's about $11,000.[6] The wood was used for just about everything: from fine furnishings and house fittings to railway lines and even sluice boxes for the gold and tin workings to the west of the Atherton Tablelands.

Logging was lawless and out of control, a wholesale system of plunder. Only the best part of the tree was cut and squared, leaving most tree limbs in a lurid trail of destruction through the forest. The cedar-getters were at the mercy of dealers who lived at the river mouths and furnished the loggers with salt, flour, tea, sugar, tobacco . . . and rum. There were reports[7] of men and women attacking grog with ferocity, then lying intoxicated on beaches—only to wake and renew their orgies until the cedar-getters' pay, such as it was, ran out and the men had to return once again to the forest.

Sitting high in water, red cedar was floated down rivers and surfed out to waiting ships. When the cedar was cut out along the coast of the Wet Tropics in the 1870s, the timber-cutters turned inland to the Atherton Tablelands where a timber rush began in the 1880s—without even waiting to see if there were buyers. As it turns out, there were none because it was impossible to transport the cut trees to the sea. But it didn't matter. In an unstoppable frenzy, by the end of 1881, 9 million super feet were stockpiled. In 1882, the stockpile was 15 million super feet.[8] (Note that 'super feet' is short for 'superficial feet', a volume measurement of a block 1 foot square by 1 inch thick.)[9] Deliriously, cedar-getters dropped 3 million super feet, thousands of trees, to be desperately pushed over the Barron Falls near Cairns and smashed to a pulp. By 1901, 16–18 million super feet, hundreds of thousands of trees, were dropped in a fever of avarice in the forests of the Upper Barron, within a stone's throw of Thiaki, and left to rot. There was no recognition of the value of anything other than red cedar. Valuable species in quantities unknown disappeared where they stood in the rush to get at and remove the red cedar.

Despite these depredations, given time, the rainforests should have healed, gradually filling their empty spaces with life. But that was not to be.

So luxurious was the rainforest that settlers thought the fertility of its soils would be inexhaustible. In pursuit of the bucolic fantasy, men and their families settled in tents or bark huts. They set to work with axes to clear patches in the forest, waiting for the dry heat before the rains of summer and then lighting blazes that would burn all but the largest logs[10]. The rainforest of the Atherton Tablelands was converted to smoke and ashes in order to establish paddocks for dairy cattle.

———

When an asteroid smashed into Earth 66 million years ago, in the shallow seas of the Gulf of Mexico near Chicxulub, on the Yucatán Peninsula, it punched a 2-kilometre hole in Earth's crust, ballistically ejecting carbonate sediment into the atmosphere at greater than escape velocity. The blast levelled trees within a radius of about 1500 kilometres;[11] kilometre-high waves crashed onto nearby shorelines, travelling up to 300 kilometres inland.[12] Tidal waves reverberated throughout every ocean basin, travelling at speeds of 150 kilometres per hour.[13] Intense radiated heat ignited wildfires across the globe. Debris, dust and gases hung in the air before falling out as acid rain. A curtain of sulphate molecules closed out the sun, causing a severe winter that lasted for decades. Simulated computer modelling shows that the global average temperature plunged to below zero for several years immediately after the impact. Australia, located near polar latitudes, was snap-frozen; ice frosted the poles.[14]

Cold ocean waters sank, churning the oceans and bringing up warmer waters loaded with nutrients. This set off a toxic algal bloom, killing three-quarters of sea life and laying waste to marine eco-systems. While computer models suggest that it took a mere 30 years for temperatures to rebound, it was too late because the damage was done. Gone were the pliosaurs and ichthyosaurs; gone were all of the non-avian dinosaurs; gone were most of the Cretaceous birds (one comprehensive study showed that, of the 20 main groups of these ancient birds, only three survived—those leading to modern birds);[15] gone were the pterodactyls. Many of the lizards and snakes slunk through, but it was touch and go: a rich fossil site in North America shows that of the 33 taxa that survived up to the end of the Cretaceous, only ten lived on beyond it.[16] The Mesozoic was slammed shut and a new era of life, the Cenozoic, dawned.

The extinction was not random. Large species were preferen-tially toppled by a global heat pulse caused by radiating re-entering material followed by the blistering cold of the impact winter. It was a killing blow for the non-avian dinosaurs.[17] Survivors once again were small and fleet,[18] hiding in the earth, in caves, in holes or under debris. Anything living in trees was doomed—surviving birds were likely ground-dwelling or aquatic.[19] In the absence of photo-synthesis, food was restricted; small insectivorous mammals—those feeding on invertebrates that ate the rotting wood and debris—were more likely to survive. In a way, dinosaurs had done mammals a favour by suppressing them. While devastated by the impact, many mammals nevertheless scraped through and proliferated rapidly into the empty niches, setting the stage for mammalian dominance for the first time since the Permian extinction of most of the synapsids.

Despite the severity of the extinction, life bounced back—particularly in places distant from the impact site, such as Thiaki. Initially, saprophytes (such as fungi) recovered on the rotting vegetation under a weak sunlight. As the atmosphere cleared, light-dependent life revived in an excited spike of activity as ferns flourished from hardy spores. The establishment of a canopy, though, took hundreds of thousands of years and comprised a low diversity of flowering plants and conifers. It was a small pool of survivors. The forests had lost their Mesozoic richness, and ecosystem recovery would take millions of years. But miraculously, 66 million years later, evolution has spun its magic once again to fill the empty spaces. Modern humans inherited an incomparable variety of life: a richness brimful of possibilities never before seen in Earth's 4.5-billion-year history.[20]

I gaze out over the Atherton Tablelands. The small village of Malanda is visible about 10 kilometres away, set among deceptively pleasant grass paddocks dotted with cattle. Sadly, I know that I'm looking at a tiny fractal corner of another billowing extinction event, one that is entirely unheralded in the history of life on Earth. According to naturalist Eric Mjöberg, a little over 100 years ago Malanda was a place 'where quite recently some new settlers had made their homes. Everywhere was virgin rainforest, the white man's axe had not yet started its destruction.'[21] Mjöberg was fascinated with the process of clearing:

> The trees usually lean in one direction, and this is taken into consideration when the trees are chopped, only halfway through. The last tree to be felled is a very large one with

a definite tilt. It falls on to the tree closest to it, by breaking off where it was chopped, and in its turn, falls on to the tree next to it, and so on . . . The rainforest is so dense and so completely interwoven with various kinds of lianas that, because of the weight, it pulls every single tree down with it. The rainforest trees are not felled in the normal way at root level, but quite a way up the trunk. This is to save the work of otherwise having to chop through the thicker diameter at the root. Therefore, one chops holes into the trunk at convenient spaces and works one's way up for several metres until a wooden board, a so called 'spring-board', is placed into it to make a platform to stand on when chopping. Experience and skill are certainly required to do this kind of work, which is also very physically demanding . . . In places, the charred trunks still lay around on the ground and here and there tall stumps stood up. In a few years' time, they would rot down and become fertile humus and the ground would be a large meadow where cattle would graze.[22]

As Mjöberg went exploring, the forest rang with the echo of striking axes when a rainforest tree was hit, gouging out one big chip after another. A forested slope was dropped, with leaves and limbs flying into the air. Out of the trees plummeted animals such as the Herbert River ringtail possum, a beautiful black possum with a white belly and a penchant for resting in epiphytic ferns. The usual catch for Mjöberg was the lemuroid ringtail possum, which he found to be 'sleeping in every tree'. 'There is really no immediate danger that these marsupials will become extinct,' he said, 'because in each large tree there are a pair of animals.'

As his trip progressed, his tone changed: 'the burnt-off rainforest area now presents a sad sight. The giants of the rainforest, previously so green and full of sap, now lie slain on the ground all wearing the sombre colour of grief after the devastating fire.' In a sanguine moment he wrote that 'the rainforest itself is necessary for their survival and in those areas being felled they are doomed to extinction'. He had seen the future.

It was 1986 when wildlife ecologists finally looked at the Atherton Tablelands, and the view was of a rainforest scrapyard. The frenzied clearing was thorough. Nearly 80,000 hectares of forest had been clawed away, leaving around 100 rainforest fragments ranging from 1 hectare to 600 hectares scattered over 1000 square kilometres.[23] Larger tracts, such as Thiaki—logged, but still intact—clung to steeper hillsides. The remaining fragments were already 20–50 years old, and they were found to be in poor health and with a dwindling fauna. Over the following 20 years, the forest mammals still showed signs of severe distress that were not stabilising, even though there was little additional deforestation.

Extinction has a cruel dynamic: small fragments below a hectare will lose their dedicated rainforest species in under nine years; medium-sized fragments of around 30 hectares will lose their mammals in three decades; and large 70-hectare fragments will lose theirs in under 60 years. At least 300 hectares is needed to keep rainforest denizens, such as the lemuroids, for a century. We would need to fully restore a third of the Atherton Tablelands to keep rainforest mammals for the next millennium.[24]

The view is of a pattern repeated in useless loops of felled forests and fires, ramified everywhere on Earth, like the endless pattern of chaos. The current Anthropocene extinction event—let's call it my extinction event—is grotesquely unique. My extinction event takes no prisoners. Common and widespread species are as susceptible to obliteration as those that are rare and restricted. Recently evolved species, such as birds, and ancient survivors, such as amphibians and turtles, are all fair game. During other extinction events, widespread species were the least likely and usually the last to become extinct. Three-quarters of extinctions after the Chicxulub impact event, for instance, involved the rarest species, often the last members of an earlier diverse group. But it is pointless to rank the current relative extinction risk of one group against another because all groups are under catastrophic duress.[25]

Refugia were crucial to the survival of life after Chicxulub. Deep valleys, rivers and lakes, coastal regions and, in the tropics, sheltered spaces offered life for birds, mammals, turtles, crocodiles, lizards and snakes. Today, all valleys, rivers and lakes, coasts and the forests of the world are fully exposed to my extinction event's shockwave. Geologically, my extinction event is happening in an instant, undoing aeons of evolution in the most rapid and self-destructive manner ever witnessed before on Earth[26]—even outdoing the extraterrestrial Chicxulub asteroid.

Since 1970, roughly 70 per cent of all individuals of vertebrate species have vanished.[27] Disaster species such as weeds have also increased by 70 per cent.[28] It's a shocking fact that calculations on global data sets of vulnerable species using fuzzy arithmetic (manipulating numbers that have a range of values) predict the potential complete extinction of vertebrates in as little as a century. This matches the fact that all of the world's forests will disappear in 100–200 years at the current rate of clearance.[29] But humanity will collapse long before this possibly, in a handful of decades.[30] Climate change is a secondary concern given the existential biodiversity crisis.

The Mabi forest that was once continuous across the Atherton Tablelands will never return. Even now the last vestiges are being scraped away, leaving only scraps—not enough to restore even a semblance of the forest's former glory. Those scraps are 'ecological zombies'—still there but not significant for ecosystem function.[31]

My gaze slides to the pert trees in our restoration plots. Thiaki survived the historic clearing onslaught, but not intact. Only 30 per cent of the forest was spared from clearing. Those patches that remain now contain severed populations of animals and plants. In many ways our restoration effort is a zombie-buster: designed to reinstate Thiaki's ecological processes, accelerating recovery by connecting Thiaki to other forests.

Unless we suddenly find the ability to take to the stars, our fate rests on efforts like this.

# CHAPTER SIX

# The walk

I eye the insect repellent—thinking of the Devonian—tuck a little notepad into my back pocket, grab the binoculars and smart device with bird app installed, and head out before six in the morning. It is cool, and a low misty cloud taunts the forest. A million leaves reach to caress the body of moisture.

My daily walk at first light starts with a jangle of chain as I open the house gate attached to an old wooden post, one of four leaning in different directions at the four-way junction near the house. The plaintive call of the endemic grey-headed robin welcomes the pre-dawn glow. This long-legged, ground-dwelling, rainforest robin heralds the early morning, the late afternoon and the evening with a gentle high-pitched piping punctuated by a *chukachuk*. The rapid-fire machine-gun-like call of the Lewin's honeyeater comes next. But it is still too early for the gregarious red-browed finches that always hang about in our front yard, bouncing, balancing, pulling on grass stalks or alighting in a line around puddles to drink. Flocks move around the paddocks, calling their high-pitched *seee-seee*.

A glorious family of king parrots is lining up in the trees in front of the house, waiting—impatiently by the sound of their querulous high whistles—for a fresh dish of bird seeds. They don't really need the food, but it's impossible to resist them.

With the house behind me, I have three choices: continue along the main ridge, on which the house sits, and take the undulating walk along the edge of the Thiaki remnant. Alternatively, I can take the path heading steeply down into the valley to my right. This crosses a little creek that was dammed for cattle and is now home to a family of small northern platypus, an evolutionarily distinct population—on the way to becoming a separate species.[1] This path then loops up and over a low ridge parallel to the main one and climbs back up to the main ridge through the gloom of the rainforest, where I could call a greeting to Tree who is nearby, before slipping along the rocky, moss-covered spring-fed creek, upstream of the platypus dam, then scrambling up a steep rainforested slope that leads to the house.

Or I could take the circuit track that slopes down to my left towards the cleared and grassed main valley in which our few fat cattle graze, then around the base of a couple of ridges and back up to the main rainforest remnant, returning to the house by walking along its dappled edge. An irritable grass-loving golden-headed cisticola buzzes and plinks at me from a perch on the fence, as I choose the circuit track and make my way down to the main valley. The east-facing side of the valley is covered with old regrowth forest that resounds with the long-whistled whip-crack call of the male eastern whipbird, the female responding with two or three chirrups. The west-facing side is patched with vigorously growing reforestation plots. I follow the valley for about half a kilometre before it plummets into a 10-metre-deep erosion gully where the creek, until now running underground, spurts clear water pleasurably as it meets the rainforested gully. The creek continues its undulating gurgle around the base of volcanic ridges and through sheer little gorges to form, just outside our boundary, Thiaki Creek—the headwater tributary of the North Johnstone River.

The track cuts around the Bull's Nose, the base of Dead Tree Ridge, which plunges down from the main ridge. When we purchased Thiaki, a lone dead tree was silhouetted against the sky on this ridge. Paddock trees like Dead Tree usually don't have a long life after their local community of trees has been removed. Ultimately, they die of stress and, I imagine, loneliness. One day a visitor, with a biblical turn of mind, said Dead Tree reminded him of the biblical burning bush. In a way that's true because today it radiates energy through a growing mantle of epiphytes, and newly planted trees crowd around its base. A fig has now claimed it as a framework.

My path skirts around the Bull's Nose just above the now rushing creek. While the water is only 20 metres below me, the rainforest hugs the banks so tightly—as if fearing an escape—that it is not possible to glimpse the creek. The elusive fernwren lives deep within this valley. Singly or in pairs, these Wet Tropic rainforest endemics hop about vigorously probing, with their tails flicking. They even squirm beneath the litter, their bills partly sealed to avoid debris blocking them. Sometimes the fernwren will follow other deep-rainforest birds, such as chowchillas and orange-footed scrub-fowls, which actively scratch among the litter and bring up the busy litter life of insects, snails and spiders. Usually shy, the fernwren's call is a drawn-out, high-pitched whistle that merges almost indistinguishably with the silvery sounds of the forest.

Above the path are more of the restoration plots. The trees are small here on the dry west-facing slope, with its shallow soils, but they are growing stolidly, and some natural recruits from the rainforest have reached out to join them as rainforest ambassadors. A small tree native to South America, *Solanum mauritianum* (tobacco bush) is also growing here. Given half a chance, this weedy species would take over in treeless paddocks if we didn't cut and poison the saplings. We have the brown cuckoo-doves to thank for that. While it's a native bird of the rainforest, the cuckoo-dove loves feeding on weeds growing in clearings or along the rainforest edge. Sometimes while we're sitting on the verandah at home, a cuckoo-dove will smash headfirst into the nearest window and eject a belly full of tobacco bush seeds. The crash victims often recover on the verandah before flying, dishevelled and with a migraine, to the nearest tree.

There are at least nine species of pigeons or doves in and around Thiaki, including the Pacific emerald dove, with its glowing green wings. Then there are the gloriously coloured fruit-doves: the rose-crowned fruit dove, with its magenta-crowned head, orange and yellow belly, and green wings; and the male of the superb fruit dove, a jewel of a bird with its deep purple head, fiery orange nape, blue band across the chest, olive-green wings and bright yellow eyes. The largest fruit-dove is the brilliantly hued wompoo fruit-dove, with its grey head, rich green back and wings, and stripe of yellow slashed across the wings. The breast and belly are plum-purple and bright yellow.

Wompoo fruit-doves always remind me of a day I call 'Fuck you Sunday', a glorious bird-coloured day of intense sky blues and deep greens, when I went out to explore the main ridge of the greater forest patch that crosses several neighbouring properties. To access this ridge, I crossed the little platypus creek, cut through the riparian forest and entered a paddock. Scraping through a fence separating our property from Barry and Kirsten's, I walked up the slope to a gate that allows access to the ridge, which had escaped clearing because of the precipitous sides dropping from it. As I walked along the ridge with its sun-speckled forest floor and sparkling leaves, I suddenly heard a loud yell, 'Fuck you', reverberating through the forest.

It was shortly after we'd bought Thiaki, and our house hadn't yet been built. I was staying in the little prefabricated cottage we had assembled for temporary accommodation. Not being familiar with the sounds of the forest—or the potentially menacing people living around it—I scanned the dark forest shadows, my heart thumping quickly with the adrenaline rush. 'Fuck you.' There it was again—closer. Deep in the forest, no one would hear me scream. I started making quick plans, deciding to bolt down the steepest slope if he came at me. It was steep and slippery, but I was fit enough to run fast while propelling myself around the slim stems. Again, I heard 'fuck you'. But this time the 'you' was a high note. I started laughing. Embarrassed at my ridiculous self, I realised the strange human-like voice wasn't a rapist but merely a bird, the wompoo fruit-dove calling *whuck woo*.

Smiling at the memory, I continue edging around the Bull's Nose as a small group of sulphur-crested cockatoos flies overhead. Birds of forests and paddocks, they've got used to my daily ramblings and no longer react in a frenzy of over-the-top panic. Flocks of screeching rainbow lorikeets fly in formation high above, straight as arrows. Along the forest edge, the acrobatic grey fantail is performing a crazy fan dance that defies gravity, twisting and turning in mid-air as it tries to catch insects.

It is early September and beginning to warm up. I spy a discarded python skin in the grass, torn and hanging in tattered pieces like an overused tissue, left by an irritably itchy snake that had clearly writhed against the ground to rip off the scrappy old garment. The snakeskin is a sign that the enigmatic echidna has emerged from burrows and can be found now, waddling along determinedly with its distinctive rolling gait, impenetrable and unfazed. In September, any female is likely to have a small, spineless puggle in her pouch after having dispensed with her suitors, who had followed her in a sexy nose-to-tail linedance until she was ready to mate. She would then have pushed up against a tree while the males dug a surrounding trench where they would butt heads, pushing each other out of the way until only one remained. In his love rut, he would have sidled up to the female and turned on his side, ready to mate.

I pull out a pen and scribble in my notebook while listening to Thiaki. Most of the gossip is about the birds. The Wet Tropics is home to around 370 species, a little less than half of Australia's birds. About 137 of Thiaki's birds are forest-dependent. Over the years, we've recorded 157 bird species in Thiaki. I'm thrilled if

I get 40 on any day. I remember to jot down the endemic lesser sooty owl I heard the night before, making its unmistakable bomb-drop whistle. Pairs of lesser sooty owls have a hundred-hectare range,[2] so there are probably only one or two pairs of these little owls in Thiaki.

Most of my bird list comprises the widespread birds, such as pied currawongs and kookaburras, whose unsubtle calls push me out of bed before dawn. The local birders don't always like the currawongs because they take the nestlings of other birds. Bold and sinister-looking scavengers, currawongs are cloaked in black and have witch-yellow eyes; at this sexually active time of year, they can have tail feathers pulled out by rivals and fly like blunt-ended Exocet missiles. This is the time of year they hold regular currawong conventions along the Upper Barron Road, where hundreds congregate to rollick and bound flagrantly about on the branches. Up at the cattle yards they monster even the kookaburras, aware no doubt of the kookaburra hatchlings sheltering in a nearby hollowed-out trunk of a tree that had its crown ripped off in 2006 during Cyclone Larry; the cyclone had twisted and churned the forest, spraying out leaves and branches like a tree mulcher.

Another regular in my notebook is the golden whistler. The male golden whistler is a spectacular bird with a bright yellow breast and white throat set off against a black collar and head. Golden whistlers breed at this time of the year, and both the male and the duller brown female attack their reflections with a persistent and annoying *rap-a-tap-tap* at the windows. Once I covered an offending reflection on the window of the shed, only to send the bird flying off huffily to the nearby ute, where it continued to attack the offending reflection in the side mirror.

Perhaps it is no surprise that a bird of such inexhaustible energy has populations that span Australia and the broad swathe of the Indo-Pacific islands. In fact, Thiaki's golden whistler, *Pachycephala pectoralis*, is a striking example of one of the most polytypic species in the world, comprising about 70 similar-looking subspecies: some have a yellow throat instead of white, or a broader black chest collar or no black chest collar at all. Although frustrating for taxonomists, this is a spectacular example of evolution on overdrive, where isolated populations on adjacent islands quickly diversify from nearby birds. The original ancestor of the golden whistler used the offshore islands in the New Guinea region as their evolutionary staging posts to the surrounding archipelagos before flowing in a wave right across the mainland of Australia.[3]

While the golden whistler may be the most polytypic species in the world, the little citrine-green silvereye, *Zosterops lateralis*, takes the cake for being among the fastest evolving of all terrestrial vertebrates. Deceptively fragile-looking creatures, with their fashionable eye-ring of white feathers, they are, in fact, the world's 'great speciators',[4] and have flown long distances to colonise more islands worldwide than any other species. A supercharged genetic mutation rate can often lead to harmful outcomes, but in silvereyes it has created super proteins that boost flight muscles which propel the birds to other islands. So rapidly are they evolving that when one group has claimed an island, the next group making landfall for a romantic interlude will be incompatible. This high evolvability[5] allows silvereyes to swiftly adapt to different environmental conditions while at the same time establishing reproductive barriers to interlopers.

Silvereyes are drifting around Thiaki now in undulating flocks, moving from one scrubby patch to the next in constant communication with a peevish *pee pee*. The flocks have thinned since the birds started pairing up. In summer, they will flock together again to waft over the landscape or, gripped by wanderlust, gather for a flying chance in the evolutionary sweepstakes that are the gleaming green archipelagos to the north, spreading from mainland Asia to the Pacific islands.

Thiaki harbours all 13 local Wet Tropics endemic birds, including the grey-headed robin, fernwren and lesser sooty owl. The other endemics are the bridled honeyeater, Macleay's honeyeater, pied monarch, Victoria's riflebird, tooth-billed bowerbird, golden bowerbird, Atherton scrubwren, mountain thornbill, Bower's shrike-thrush and chowchilla.[6]

The bridled honeyeater is a medium-sized brown-grey bird, washed with olive, and with a yellow bridle extending from its bill to its ears. Active and aggressive, it loves nectar and will call loudly with a rather wheezy downward descending *tshew-tshew-tshew*. Quieter than the bossy bridled honeyeater is the Macleay's honeyeater. Only an occasional visitor from lower altitudes, this bird prefers to probe for insects and spiders in tangles of leaves, high vines, rotting wood and crevices, and it buries itself halfway into certain flowers searching for nectar. The pied monarch also prefers lower altitudes, and I've glimpsed it only once deep within the forest. Victoria's riflebird and tooth-billed bowerbirds are commonly heard at this time of the year, since it is their breeding season. The male tooth-billed bowerbirds are spread out near their bowers along the main ridge, within hearing distance of each other. The bowers are simple affairs—just a swept circular patch of forest between tree buttresses—but they are strewn with leaves that have grey undersides and, when placed upside down, glow in the gloom. The leaves are cut neatly with the double notched—or toothed—bill.

It was years before I saw my first golden bowerbird. I was returning from another circuit that takes me down the main road for a few kilometres before I cut across a friendly neighbour's paddock and plunge into the forest and across the creek that begins in our main valley. Here the water tumbles over a little waterfall into a small pool. From the pool it is a slog up another steep ridge along an old and overgrown forestry track before I finally emerge at the edge of one of Thiaki's paddocks. It was here at the forest edge that, one day, I heard what sounded for all the world like an old Massey Ferguson tractor coming from an overhead branch. Disoriented for a moment, I glanced up and caught my breath—there was not one but two male golden bowerbirds backlit on a low branch, almost translucently yellow. They were sunshine incarnate.

Golden bowerbirds love cool upland forests at elevations from 700 to 1500 metres. As temperatures rise, they move upwards and their habitat area decreases. Currently, with 1-degree Celsius warming, the golden bowerbird habitat has already decreased by half to 582 square kilometres. At 2 degrees Celsius, their habitat will be reduced to 163 square kilometres, and they will have disappeared from Thiaki in their upward migration. With 3-degrees Celsius warming, their sunshine will fade away as they run out of altitude and become extinct.[7]

Back at the Bull's Nose, the Atherton scrubwrens are encouraging me to piss off with raspy little *tss-tsss*. Like little mice bouncing around the lower part of the canopy, they are the most secretive of the four types of scrubwrens. Small, insectivorous, dun-plumaged birds, scrubwrens can be intolerably difficult to tell apart particularly if, like me, you tend to use birdsong for identification. The Atherton scrubwren's *chee-wip* is very similar to the *cheeWIP* of the large-billed scrubwren, which in turn is close to the slightly higher pitched and more rollicking *ch-weep* of the white-browed scrubwren. More or less on the same theme is the *chzweep-chip-siep* of the yellow-throated scrubwren. It intersperses its call with a regular sharp *tik*, which is frustratingly similar to the *tiek* of the white-browed scrubwren. Another little brown bird, the endemic mountain thornbill, has an unmistakably big call of whistles spliced with more melodious notes. It is one of the prettiest calls in the forest, though the bird is usually chasing competitors angrily when the call is the most appealing. Groups of mountain thornbills hop and bop about in Thiaki's mid-canopy, flitting from tree to tree with characteristic undulating flight.

Given the similarity of the calls, it's no surprise that the scrubwrens are closely related. Three of them are of the same genus, *Sericornis*, with the Atherton scrubwren, *S. keri*, most closely related to the white-browed scrubwren, *S. frontalis*. And both of these are closely related to the large-billed scrubwren, *S. magnirostra*. The yellow-throated scrubwren, *Neosericornis citreogularis*, is only distantly related but is closer to the scrubwrens' common ancestor. This bird lived in the wet forests of northern Australia about 13 million years ago, at around the time the island of New Guinea was being squeezed up along Australia's northern bumper in a complex collision that also collected a scattering of island-arcs. The patchwork of habitats in a drying Australia—together with the new empty niches emerging from the sea as New Guinea—led to about 19 different species spawning from this one Australian ancestor.[8]

Bower's shrike-thrush—a slate grey bird with a buff breast and black beak—is one of my favourite birds. It has a large repertoire of summery calls, from a beautifully melodious call interspersed with *chuks*, to a loud *wee-jaw, wee-jaw, wee-jaw, quort, quort, quort*. So attractive are the calls of Bower's shrike-thrush that the tooth-billed bowerbird has appropriated them as favourites when calling from a bower.

Not far away, I can hear a family of endemic chowchillas jamming in the rainforest. I was introduced to them in 2005, when I first stayed overnight at Thiaki. Outside of the prefabricated hut, I heard on the gentle breeze a syncopated jazz riff in three/four time. Chowchillas are little dark olive logrunners with a sassy pale eye-ring. They hang out in cooperative family groups, cheerily sharing any insectivorous food caches found. Each morning they warm up with 30–60 minutes of scat singing, each member taking turns. Recordings of songs have identified 56 different song elements,[9] but there will be many more because some warbles are simply too low in frequency for human ears to hear. Chowchilla improvisations have exceptional regional variation; chowchilla groups carefully listen to neighbouring groups' songs and, like any musician, adopt new elements to enhance their repertoire. The group grooving around the Bull's Nose where I stood listening is noticeably different from the one at the homestead. While scientific explanations for the variation have centred on territorial defence and reproduction, it is clear to me that they simply love to jam.

My attention is drawn to a busy white-throated treecreeper. This inoffensive and inauspicious little bird, bounding up and around tree trunks, hunting for insects and issuing a penetrating piping call, is a direct descendant of one of the very first of Earth's song-sters.[10] Fifty-five million years ago, these birds—together with the ancestors of lyrebirds—inhabited the fresh new forests of the Cenozoic, enlivening the post-Chicxulub world with song.

The extraordinary thing about all of these birds—in fact, about all birds—is that they don't have teeth. Birds are dinosaurs, and dinosaurs were so toothsome that dinosaur teeth have piled up in banks of sediment dumped by Cretaceous rivers.[11] Birds emerged seamlessly from theropod dinosaurs, many of whom looked just like birds—feathers, wishbones, air sacs and all. Any Mesozoic birder would have struggled to tell the difference between a *Velociraptor*-type animal and an *Archaeopteryx*-type animal.[12] When the Chicxulub asteroid ploughed into Earth, the planet was packed with toothy bird-like dinosaurs—and all birds with teeth died alongside *Tyranno-saurus* and *Triceratops*. Only the toothless neornithines survived. These ground-dwelling birds must have been common and looked similar to their toothed relatives except for the presence of beaks, which they used to process a specialised diet such as seeds.[13] After the Cretaceous Armageddon, seed banks may have persisted for upwards of 50 years among the smouldering detritus as they do today after clearing and fire. And it was this fortuitous diet that got them through.

———

My dentist advised that I brush with an electric toothbrush, so I bought a battery-powered one that can run for days to take on trips out to the bush. Unfortunately, it has an irritating ring of light like a micro flying saucer that shines up into your eyes, obscur-ing vision at night. It also has annoying built-in sound effects so you know which gear you're in. This, it seems, was all too much for a female howler monkey, which dropped down to the nearest low branch to peer incredulously at me and the bizarre extra-terrestrial creature I was shoving into my mouth. The next evening, she brought the whole family, including a toddler clinging to her back. The family ran a nonstop chattering commentary during the procedure. I sent my battery flat continually restarting the gadget to converse, albeit in this bizarre fashion, with the comical family.

I was at Cocha Cashu, an Amazonian research station in one of the remotest national parks in the world. Manú National Park is a huge 1.5-million-hectare park that spans the Peruvian Andes to the lowland Amazon jungle. It is one of the hottest of global bio-diversity hotspots, and I was fortunate to be invited by colleagues to visit there in 2018. The place is full—every nook and cranny, every niche, is occupied by life. There are more than 200 species of mammals, 800 species of birds, about 150 species of reptiles and amphibians, 1300 recorded species of butterflies from hundreds of thousands of invertebrates, and thousands of plant species.[14] Its vertiginous variety stuns even seasoned researchers, who continue to discover new species.

How can one describe a howler monkey morning chorus? More felt than heard, it is a roaring and pulsating wave of sound that vibrates every cell in your body. At up to 140 decibels, it is liter-ally deafening. While howler monkeys might be one of the loudest animals in the world, for mystery and magic nothing beats the tremulous and mournful piping of the hen-like tinamous. And this song, it turns out, is an echo of the music of the Mesozoic.

Several species of the shy and secretive tinamous are found in Manú. Cryptically coloured, compact, ground-dwelling—but not flightless—omnivorous birds, they poke around with their bills to feed on seeds, fruits, insects and small vertebrates. It is a lifestyle they inherited from their ancestor, a hen-sized bird called *Lithornis* that lived in the Cretaceous of North America. And, as unbelievable as it sounds, it is from this bird that the ratites—including the emu and cassowary—evolved.

Ironically, ratites found their evolutionary niche by abandoning the very essence of 'birdness'—the ability to fly[15]—even though it was flight that was at the very heart of becoming a ratite.

Charles Darwin predicted that all ratites—cassowaries, African ostriches, Australian emus, New Zealand kiwis and South American rheas, as well as the now-extinct New Zealand moas and Madagascan elephant birds—were related. But how did they end up spread across the southern hemisphere? It was Darwin's contemporary, Thomas Huxley, who observed that the tinamous had the same reptilian arrangement of bones in the roof of their mouths as ratites. This stamped the chubby tinamous as ratites.[16]

It took a century and a half to come up with a well-founded explanation for the distribution of ratites.[17] In 2017, a team of scientists published a paper explaining that they had managed to get their hands on some DNA from the subfossil bones of recently extinct elephant birds. Despite the fact that these birds weighed a whopping tonne—making them the world's largest bird—their DNA revealed that their ancestor was the small hen-sized *Lithornis*. This begins to explain the fact that the closest relative of the elephant bird is the 2-kilogram kiwi of New Zealand. *Lithornis*, a denizen of the northern hemisphere, sits at the base of both family trees, and also at the base of the ostrich lineage. This lineage started evolving about 80 million years ago in the northern hemisphere after *Lithornis* crossed to Africa when that continent docked against Eurasia.

Between 70 and 80 million years ago, *Lithornis* also fluttered across to South America, probably via the inter-American land bridge. There they evolved into ratites and tinamous. Early ratites—just like some birds today—loved to colonise islands, and they flew to Madagascar and New Zealand, which were then bunched closer to Antarctica with Australia. This is why the kiwis and elephant birds are close relatives. *Lithornis* took cover from the impact of the Chicxulub asteroid before flying on to Australia,[18] where they found an entire continent emptied of sizeable predators.

Birds that colonise remote islands are released from the natural-selection pressures of predatorial continents. Most strikingly, they lose the need to fly. And they can do this in as little as 20,000 years. It seems that it is costly to keep the strong muscle mass of flying forelimbs, and this muscle is instead moved to support larger hind-limbs. Rails, parrots, pigeons, owls, waterfowl and passerines all do it.[19] In Australia, natural selection moulded *Lithornis* into the emu and cassowary.

———

I'm brought out of my reverie by the scrambling clubmoss, *Lycopodiella cernua*, growing in a shaded niche tucked into the steep bank of the path around the Bull's Nose. It reminds me not only of the deep time roots of Thiaki but also of the resilience and eccentricities of life. This plant is a survivor from the Devonian of 380 million years ago, having come through several major mass-extinction events to be one of the most widespread of the club mosses and found across most tropical areas of the planet.

Once around the Bull's Nose, the path climbs steeply up the side of a ridge paralleling Dead Tree Ridge. Two small valleys, falling on either side of this ridge, are now filling up with reforestation trees that themselves act as a framework for new recruits from the rainforest. The track meets with the main Thiaki remnant at a spot we call The Three-ways, a place where I saw my first cassowary in 2017. Only about a year old, it looked like a large shaggy feather duster standing at the edge of the forest. It had been banished from its dad's care (the mum lays, and then promptly abandons, the eggs) to lead a mostly solitary lifestyle. The young brown bird would hopefully grow into an adult bird with glossy, black, hair-like feathers. The neck and head would be flooded with brilliant blue and purple at the front; the back of the neck would be pink; and the neck would be adorned with long, drooping, red wattles. Standing on legs that are long and strong with three toes, the youngster is armed with a small dagger-shaped claw, known to disembowel, attached to the inside toe. It was impossible to tell the sex of this youngster. Females grow tall, almost 2 metres, and weigh over 70 kilograms. Males are slightly smaller.

A tall helmet that looks like an Egyptian head ornament will crown the bird as it grows with age. Known as a casque, the head ornament has attracted scientific curiosity for 200 years, some speculating a protective role while the bird is bolting at high speed through the forest with its neck outstretched. Others have suggested it might be used as a resonance chamber to amplify their low-frequency booming call. Recent research, however, suggests another role.[20] Uninsulated and highly vascularised, the casque is more likely a thermal radiator. Heat regulation is particularly important for an animal of such large size and dark colouring in tropical habitats— try doing field work in the tropics wearing black. Cassowaries are known to put their heads in water when it is hot to aid the air-conditioning effect. And when the birds are cold, circulation to the casque is reduced to restrict warm blood to the core of the body.

This juvenile cassowary—with skin colour, black feathers and casque still to come—checked me out, gazing out of amber eyes, rumbled like an approaching truck, and then sauntered with teenage attitude into the forest depths, clearly not as excited as I was. I'd heard that a cassowary was in the area, and I'd wished and hoped that this icon of Australia's rainforests would come here to check on our reforestation efforts, which included plant species with fleshy fruits that cassowaries are known to enjoy.[21]

While cassowaries rely on the fleshy fruits of over 240 species that have fallen to the forest floor, forest plants in turn rely on casso-waries to distribute their seeds. Cassowaries are the only Australian animal capable of moving large fruits of rainforest plants en masse and by up to 5 kilometres.[22] Since most of Thiaki's forest is fleshy-fruited, the cassowary is instrumental in shaping the very structure of the forest. And it does this simply by pooing. Cassowary dung is an unmistakable porridge of large fruits and seeds, and a single cassowary splat can weigh 6 kilograms or ten per cent of the bird's weight.[23] The cassowary's digestive tract is quite short, so the fruits only spend a short amount of time travelling through. It seems that the gentle processing improves germination because if you come across a pat a few weeks after it has been deposited, it will have germinated into a mini forest plantation.

Cassowaries are locally called dinosaur birds. This tautology underscores the fact that the bird—as it appears today—would not look out of place in the Mesozoic, mingling with other crested dinosaurs and pterosaurs. The distant ancestral cassowary was, of course, the resilient *Lithornis*, which must have flown through a global holocaust, fleeing across white-capped polar Gondwanan seas before landing on a continent devoid of predators and with relaxed evolutionary pressure to eventually morph into a giant. It's a mind-boggling ancestral story full of evolutionary derring-do.

Today, however, even these indomitable survivors are endangered because the very forest floor is being ripped up from under their sturdy feet. The southern Atherton Tablelands, where Thiaki resides, is one of the eight key areas identified in the 2007 Cassowary Recovery Plan where cassowaries are under the greatest threat due to clearing and fragmentation.[24]

Tragically, a few days after the cassowary's visit to Thiaki, the teenager was caught on a neighbour's property. It was seen to be a nuisance, bundled up by the Queensland Department of Environment and Science, and sent to a nearby cassowary-rehabilitation centre, where it died.

# CHAPTER SEVEN

# Mabi's world

In the rainbow,
In the shapes within mirages,
In the spray of a waterfall,
Living flames in camp fires.

In the infinite depths of a waterhole,
The misty breath of mountain and trees,
Thiaki is, and has always been, steeped in spirits.

Penny van Oosterzee

The Aboriginal perception of ancestral time, according to Deborah Bird Rose, is ecology on steroids: life flourishing through looped and tangled benefits mediated through earth, water and air, a potent power, with us always, sustaining and maintaining the world, an ever-present condition of existence with no limitation of space or time.[1] Uncannily like evolution, the description is deeply philosophical: we, alongside all living forms, are part of the process with a past time reference that is played out in the current time.[2]

This totality of interactions creates life on Earth. But this cosmology comes with a sting: if we fail it, it will fail us.

———

From the main Thiaki remnant, the view to the north and east is of rolling emerald fields and crops, dotted here and there with solitary paddock trees, shadowed by dark patches of rainforest scraps and small national parks. Tendrils of creek lines sneak through a grassy matrix. Tourists often stop near our driveway to take a picture of the bucolic outlook across the Atherton Tablelands.

But fallen with the trees from this picture-perfect view are the rainforest peoples. The Wet Tropics is the home of at least 20 Aboriginal groups, 120 clans and around eight languages.[3] The Ngadjon people who inhabited Thiaki had mind-mapped and named every conceivable aspect of the place. Every boulder, waterfall, cluster of trees and individual tree flange, gorge, valley, river stretch, riffle, ridge and slope would have had a story associated with a spirit or ancestor. More than scenery, it is an inhabited geography: each feature detailing ancestral adventures. Luminous and numinous, Country remains for many a sentient landscape: 'the waterholes, the increase sites for plants, animals and humans, including those generations of deceased people and unborn people, the fishing places and yam grounds, the birth and burial sites'[4] and all of the Dreaming sites and tracks as well as the songs and dances that channel and direct the power of the Dreaming—invoking the living and the dead—are all there, pulsing through people's daily lives.

I imagine that travelling in this sentient, spirit-saturated place would have been like travelling over a beloved's body. To leave such a place, in the ordinary course of events, would be unthinkable.

'In this unfolding view of fields and farmhouses, rainforest pockets, and rural townships,' wrote Sandra Pannell in her historical topography, *Yamani Country*, of the Atherton Tablelands, 'the contours and coils of *Yamani* [the rainbow serpent] can still be glimpsed.'⁵ Even stripped of its waves of rainforest, the surviving Ngadjon see not only the 'smooth planisphere of paddocks, fence lines and bitumen roads' but also 'a dense forest of Aboriginal memories and European names, supporting a tangled undergrowth of deeply personal experiences enmeshed in broader historical events'.⁶ Simultaneously ecological, social and cosmological, it has a dark underbelly.

———

At the base of Mount Bartle Frere is a place the Ngadjon people call Top Camp, 20 kilometres east of Thiaki. There, enveloped in mist created by mountain spirits,⁷ a group of Ngadjon women once sang with faces raised to the sacred place on the mountain where the spirits resided: where the women, too, would journey when they died.⁸ As the mist lifted, the women would shift their gaze to the Atherton Tablelands. Beneath the bright green fields of cows, and crops of tea, another landscape shimmered. There, just below a farmhouse, is a crater, known by the Ngadjon as Djilan. It is where Yamani used to reside. He lay warm against a fire, but the old people, in the form of birds, wanted the fire because they were cold.

> Where that waterhole is, where that swamp there is now, years and years ago, back in the old Dreaming, it used to be real cold, and lot of storms and lightning must have hit a log and lit the log up for fire. All these birds were all cold sitting on the edge of the swamp. They see that big Yamani there and he's all warm. They all shivering because they cold. They used to fly down there and try and get a bit of fire-stick to get warm and every time they did he would chase them away. And this Spotted Drongo came there one time. All the birds reckoned they were cold, and told him that the Yamani wouldn't let them grab the fire, and he said wait, wait until

the Yamani asleep. Spotted Drongo grabbed that fire-stick, and as he was coming out of the swamp the Yamani woken up and got wild with him. He flipped his tail and when he did this he hit the back of the bird's tail and split it. That's how come got that fork in that tail of the Spotted Drongo. He managed to get up to the top with that fire-stick and all the birds were singing out, real happy they got that fire-stick. That Yamani got the shits with them mob, with those birds then and that is when he moved to Lake Barrine. Early in the morning. They reckon if you see it from an aeroplane you can see that the trail of the lava flow, where he come all the way to here. He went through the earth and made that track, and the water follow him here, and he bin come here, to Lake Barrine, he left early in the morning. *Barriny*, 'early morning'. When those birds woke up that water was gone, but they didn't worry because they all warm. They got their fire-stick. *Barriny*, that's early morning, they reckon this lake going to move again. He was supposed to move from that road over there but they cemented it all down there and put the wall there. Nan and that reckon he's going to break that down and go.[9]

A friend, a geologist who was training to be a pilot, confirmed for me on a training flight that you could indeed see Yamani's track, a line of rock from the crater—near where the women sang—to Lake Barrine.[10]

———

Dressed in a neat, ironed, white shirt and a grey vest, dark long pants and runners, the senior man of the Ngadjon people walked briskly across the road, leaning lightly on his walking stick, heading for the butcher's shop. 'There's Uncle Ernie,' said a friend with whom I'd been chatting outside the local supermarket. It was uncanny.

I had just been asking about the Traditional Owners of Thiaki and was told that Uncle Ernie was the best person to talk to. And there he was. My friend waved to him as he came out of the butcher's shop, in the hope that he would stop for a chat, and he waved his walking stick in brisk salute and pointed it towards the car park. Here, among the cars outside the Spar Supermarket, I introduced myself to Ernie Raymont, who had an intimate—more or less exclusive—relationship with Thiaki.

Ernie lived with his sister, Margaret, in a tidy brick house in Malanda. (Since writing this book, Margaret has sadly passed away.) One day around the kitchen table, Margaret told me that while the ground of Lake Barrine belongs to the neighbouring Yindinji people, the water is the responsibility of the Ngadjon because it comes from Ngadjon Country. I'm overwhelmed at the finely divided responsibility of every aspect of the landscape. Margaret's totem was Djilan.

Djilan, behind the milking shed of a dairy farm, was an important ceremonial ground. It was also the site of a recent massacre. Here, around the turn of the twentieth century—as retribution for the murder of miners who were raping Ngadjon women—a clan of Ngadjon men, women and children was mustered by the police and shot. Ernie's grandmother, Molly Raymond, told of her mother finding her husband hung up in a tree, still alive, trying to push his stomach back in. The police, she said, were around all the time, hunting and killing people in the most gruesome manner, such as dragging girls through fire before killing them and cutting them up.[11]

———

Creation stories and science can bleed into each other.

The Ngadjon say that the three crater lakes—Yijam (Lake Eacham), Barriny (Lake Barrine) and Ngimun (Lake Euramoo)[12]— are sentient entities who had come from a long way away to settle in the Atherton Tablelands. One day, according to legend, everyone was out of camp save two newly initiated men who were not allowed to leave, not even to defecate, without an escort. But needing to shit they called without success for an escort and then went out on their own. Against the strictest of taboos, they saw and tried to spear a wallaby but missed and hit a sacred tree. When they pulled out the spear, it had a grub on its tip. They returned in shame to the camp, as the sky began flaring red and black.

As the others returned to the camp, they saw the grub on the end of the spear and demanded to know what the two young men had done. The earth under the camp suddenly began to shake, twist and crack, roaring as water gushed to the surface. The wind howled as if a cyclone was coming. The ground then opened up, swallowed all the people and formed the three lakes.

Yijam, who had erupted first to consume the people, spat at the other lake, Barriny to stay put, but she refused. In the meantime, Ngimun had begun to erupt. Barriny came up spewing at daybreak, which is why her name means 'dawn coming'. Radiocarbon dating of sediments overlying the volcanic material—with Lake Eacham dating to about 9000 years ago, Lake Euramoo to 11,000 years ago, and Lake Barrine to 17,000 years ago[13]—confirm the order of the eruptions.

Our home sits atop the three-million-year-old Malanda volcano. Splaying out from the summit are its eroded flanks and valleys. On mornings when the Atherton Tablelands are blanketed by swirling mist, it's easy to imagine it as a sea of lava with only the tips of single-crater cinder cones emerging from the rising drifts of misty breath curling up along their flanks.

Geologically, the Atherton Tablelands are known as the Atherton Basalt Province, and it comprises 65 eruptive centres dating from seven million years ago to less than 7000 years ago.[14] The Malanda shield volcano is the largest in the province and, around three million years ago, it exploded in a voluminous gush of lava that filled valleys up to a depth of 120 metres, oozed down the ancestral Johnstone River valley and poured over the Great Escarpment and onto the coastal plain. I imagine that it was something like the Kilauea shield volcano in Hawaii, which erupted in 2018[15] with fast-moving lava flows and lava fountains spewing up to 80 metres high; it was accompanied by sky-eating sulphuric mushroom clouds 8000 metres high, explosions and earthquake swarms exceeding 700 per day at magnitude 4, ground collapse and a hail of tephra.

Lake Eacham, one of the youngest volcanic features on the Atherton Tablelands, is a maar volcano. These erupt even more explosively than shield volcanoes because ground water comes into explosive contact with the lava, creating earthquakes and howling wind. Maar volcanoes characteristically fill with water and become crater lakes. The Ngadjon believe that the Yamani inhabiting Lake Eacham has been there a long time, so may be sick of the place and be ready to move again.[16] Strictly speaking, there's no reason why this couldn't happen since the area—with its deep history of volcanic activity—is still active.

———

From our vantage point atop the Malanda volcano, the scene of a massive tectonic battle is spread before me. To the west is the Herberton Range—remnants of the mountains that squeezed Thiaki from the sea 360 million years ago—part of the spine of the Great Dividing Range, tapering as it approaches its northern extremity in Cape York Peninsula. The Barron River rises on the eastern slopes of the Herberton Range and collects water from the western and northern Atherton Tablelands.

To the north-east is the rugged Lamb Range. On the eastern horizon is a sheer escarpment with the formidable silhouetted peaks of Bartle Frere (1622 metres) and Bellenden Ker (1593 metres), the highest mountains in the state, both within the Wet Tropics of Queensland World Heritage site. The Mulgrave River and Russell River drain these sharp-edged slopes. The ranges and sharp escarpment were formed about 100 million years ago when the

Pacific Plate—which had been plunging beneath Australia, dragging down the continental crust—slowed and stopped; the action suddenly shifted eastward to where New Zealand was tearing away from Australia. Released from the effect of the plunging plate, the continental crust under eastern Australia rebounded, tipping out the Cretaceous seas and creating the great escarpment of Australia's eastern coast.[17]

Where the rupture with New Zealand was occurring, the land stretched like a baking cake rising in the oven,[18] cracking into angular blocks under the strain. Some blocks sank to form the ocean floor. Others were thrust high and dry, forming an acute escarpment rising in sheer cliffs from the ocean and running the length of Australia's east coast. Tumbling waterfalls and rivers etched the margins into retreating edges that wandered along an emerging coastal plain on which towns, such as Cairns, now lie. The Atherton Tablelands itself is a recent volcanic tongue—pimpled with cinder cones—lying between the Herberton Range and the coastal escarpment.

When Australia finally tore away from Antarctica and moved towards the equator, north-eastern Australia began to ride over a large lava mantle plume that had been stirred up as a result of New Zealand's divorce from Australia. Deep wells plumbed into the mantle plume gave rise to strings of 'central' volcanoes that mark the continent's traverse over the plume: from 34-million-year-old Cape Hillsborough in central Queensland to six-million-year-old Mount Macedon in Victoria.[19] Lava-field eruptions, such as those underlying the Atherton Tablelands, however, were shallower. These wells tapped into smaller pools of lava caught up in the shearing and smearing forces between the movement of Earth's crust and the mantle, particularly in places where the mantle was in active motion.

Thiaki sits adjacent to arguably the most tectonically active region in the world.[20] It is where the Australian and Pacific tectonic plates continue to collide and are now throwing up New Guinea and squeezing Indonesia up between Australia and Asia. No wonder northern Australia is occasionally stirred into volcanic activity.

———

Ernie Raymont's totem is Lumholtz's tree-kangaroo, known as *Mabi* by the Ngadjon people.

> *Mabi* is my totem. I'm not allowed to eat the tree-climbing kangaroo. It's my totem. He's my countryman. I'm not allowed to eat my countryman. They've started to move out too, with the warming in the climate. They've gone over to the mountain ranges over there, near the Crater [Mount Hypipamee]. It's a lot higher and a lot cooler for them there too. So they are getting away from the lowlands to the highlands where it's a lot cooler. Most of the dairy farms up here on the Tableland still got patches of rainforest on their property. So they go from one rainforest to the other. People don't know this is going on.[21]

This may explain why we, just a couple of kilometres from Hypipamee, have such a large population of the tree kangaroos. I see them or hear their bark almost daily. Often enough, as I work at my desk next to a window looking out into the rainforest, I will have the company of a tree-kangaroo metres away, curled on a branch, its long tail swaying with the movement of the trees.

Tragically, Ngadjon creation stories were mostly swept away, victims of the Australian Wars. Perhaps terrified of deadly retribution, Ernie's grandmother, Molly, the last Ngadjon person to be born in the forest, at Top Camp, was reluctant even to speak her language. Ernie explains that when he was a child, sometimes he'd hear her talking with the other older people in their language, but as soon as the kids came near, they'd quickly revert back to English.[22]

Ernie was born at Mona Mona Aboriginal Mission near Kuranda, in 1942, where his father, grandfather and grandmother, Molly, had forcibly been sent after an altercation with the Malanda Police over the use of opium. The family was moved back to Malanda when they received an exemption from provisions of the 1897 *Aboriginals Protection and Restriction of the Sale of Opium Act*, which controlled the movement of Aboriginal people well into the twentieth century. Molly built a traditional dome *midja*, and the family lived in it for a short while before moving to a nearby settler family's property. The pioneering English family employed Molly in the Malanda Pub, and Ernie went to work on cattle stations in western Queensland.

The Mabi creation story centred on Jordan Creek, at the base of Mount Bartle Frere, was never passed on to Ernie. 'They never told me how it was created, how Mabi came to be in this country and how it came to be my totem. You'll find that with a lot of people, they'll know their totem, but they don't know the story behind it. I'm a bit sad about it.'

———

The science story of Mabi and its ancestral world—while a poor substitute—is nevertheless a complex and fantastical one involving a global marathon battling brutal climate change, both blistering and freezing, an earth-shattering meteorite, the planetary influences of Jupiter and Saturn, super volcanoes and predatory monster dinosaurs, toothsome birds and mammalian carnivores.[23] It was an ordeal like no other.

According to science, Mabi is a seasoned survivor of an evolutionary journey beginning around 125 million years ago in China, where the oldest ancestor of marsupials, *Sinodelphys szalayi*, evolved. Remarkably, an entire fossilised animal has been found with a surrounding halo of fur, so we know it was a 15-centimetre mammal that weighed perhaps 30 grams. These first marsupials were petite and agile climbers capable of grasping and branch-walking, and they were probably active both in the trees and on the ground.[24] From China they scampered eastward, braving the Bering Land Bridge, and westward through Europe where they left evidence of a romance, in France, in 95-million-year-old descendants. From

Europe it was a straightforward trip to North America, chunks of which loitered nearby in the post-Pangaea world, and from there to South America. Over the next 20 million years, marsupials evolved into 22 different families—many more than exist today.

While it would have been difficult to tell the difference between placental and marsupial mammals, which were both in abundance in North America close to the Chicxulub impact site, for some unknown metabolic reason the impact hit the marsupials the hardest. Luckily, those that had migrated to South America escaped the full brunt. They continued to move south, eventually finding themselves on the Antarctic Peninsula that, 64 million years ago, was a forested archipelagic extension of South America.

Unlike today, Antarctic temperatures were moderate, with warmer months reaching a balmy 25 degrees Celsius; during the polar winters, the temperature dropped to a relatively mild 2 degrees Celsius. In these conditions, life in western Antarctica flourished: fossil evidence reveals at least ten species of primitive marsupials, as well as a weird and wonderful menagerie of mammal-like gond-wanatherians; ancient placental Xenarthra, related to tree sloths and anteaters; Litopterna, a South American ungulate; hoofed Astra-potheria; and a dryolestid species that seems most closely related to the last common ancestor of mammals before placentals and marsupials split.[25] Wanderlust and empty niches drew some mammals deeper into precarious polar latitudes to probe the wilds between western and eastern Antarctica, which hovered over the South Pole.

Not only was it colder here, but tectonic forces were also pushing up the Transantarctic Mountains. By 55 million years ago, they soared to 6 kilometres, becoming one of the most formidable terrestrial barriers on Earth. Any animal slogging its way towards Australia through the tall coniferous and Antarctic beech forests would have had to brave polar temperatures that, while higher than today, were still a chilly –5 degrees Celsius. Metabolically, these animals would have been small and fleet; they would have had the ability to climb trees and to hibernate or use torpor to conserve energy—something nearly half of today's Australian marsupials still do.[26]

Those who emerged from the looming Transantarctic Mountains were confronted not by the promised land but by a choppy seaway. Straining to begin its own journey,[27] Australia had ripped open a vast ocean-filled rift all the way along its southern coast from the west coast to Tasmania, leaving only a scrap of land called the Tasmanian Gateway to stumble across.

———

Mabi's early world can be glimpsed in a hole in the ground in a farmer's paddock near Murgon, south-east Queensland. Here at the Tingamarra fossil site, palaeontologists shovel tonnes of green-brown clayey ooze into bags, which they take back to the laboratory to wash, sieve and sort, hoping to get a glimpse of life in the Eocene of 50 million years ago. Out of the muck might appear a few lost teeth, a small jaw with a tooth in it, or tiny leg and ankle bones. Seemingly unspectacular, they are, in fact, revolutionary. They provide a flickering glimmer like a candle throwing shifting shadows across a 90-million-year Australian dark age, from the time of the strange monotremes and primitive mammals of the late Cretaceous to the sparkling whole skeletons and skulls of the 25-million-year-old rainforest world of Riversleigh in northern Australia.

Home to a motley group of ancestral waifs, Tingamarra has produced the oldest Australian marsupial, *Djarthia murgonensis*, which, going from its fossilised abundance, dominated the Eocene environment, nosing through litter and scurrying about in the trees. In its ceaseless activity, *D. murgonensis* seems to have scuttled right back to South America, where it evolved into a group of animals called microbiotheres. Only one of these still lives today—a finger-sized bush monkey called *Dromiciops gliroides*. At Tingamarra, the affinities with South America are also found in fossils of three shrew-like marsupials. There is also evidence of the oldest bat in

104

the southern hemisphere, *Australonycteris clarkae*, the oldest snakes and the oldest frogs.[28]

It is an extraordinary suite of ancestors, only outdone by the intriguing and unexpected discovery of the first placental mammal to set foot on Australia. This rodent-sized insect- and fruit-eating placental, *Tingamarra porterorum*, must have come in the same migratory wave with the marsupials, ahead of the looming Trans-antarctic Mountains. Its discovery overturns, in one fell swoop, the idea that placentals will always out-compete marsupials.[29] Missing, however, is the ancestor of Ernie's tree-kangaroo country-man, even though, according to molecular modelling, this animal was already living.

One set of fossils from Tingamarra provides a key of another kind—tiny ankle bones proved to be the oldest fossils yet known of passerines, or songbirds.[30] With about 60 per cent of all birds today being songsters, the world is dominated by them. Up until the end of the twentieth century, it was thought that Australia, an antip-odean outpost, received its birds in waves from more evolutionary advanced Eurasia even though there was no evidence for the idea. Australian birds were even given European names such as warblers, robins and creepers. In the 1980s, however, this idea was turned on its head by genetic work that showed songbirds originated in Australia.[31] The tiny fossil songbird ankle at Tingamarra provides this idea with more than a leg to stand on.

Indeed, it appears that Western Gondwana (South America, West Antarctica and parts of East Antarctica) was the birthplace not only of songbirds but of *all* modern birds.[32] Except ratites. And even these passed through Western Gondwana in the guise of *Lithornis*. It seems that the toothless birds that survived the Chicxulub asteroid quickly moved into the empty niches, evolving into all of the modern types of birds: freshwater and marine birds, predatory and noctur-nal birds, ground-dwelling birds and arboreal birds. From Western Gondwana, some flew north through the warm and wet climates of South America to the northern hemisphere. Others were drawn into the cool and temperate climates of Antarctica, before continuing to Australia. These include the early ancestors not only of songbirds and parrots, but also of fowl, nightjars, frogmouths, pigeons and some waterbirds.

———

Across the fat belly of Earth, other ancestors burst onto the scene in the wet and hot climates that marked the Eocene:[33] primates. Resembling tarsiers with big eyes, a shortened snout, grasping hands and feet tipped with nails rather than claws, they leapt through trees of the northern continents of the world.[34]

Like the animals of Tingamarra, primates flourished through a brief interval of sky-rocketing global temperatures caused by a massive pulse of volcanically produced carbon dioxide, which sent temperatures soaring up to 8 degrees Celsius higher than today, triggering a release of methane from ocean sediments and from permafrost at the poles. Fast and furious, the event lasted less than 20,000 years, but it heated the planet for the next 200,000 years. Polar seas reached an astonishing 23 degrees Celsius, causing massive extinctions in marine environments. On land, however, the forests that cloaked Earth from pole to pole mopped up the carbon dioxide, while sediments produced by heavy rainfall trapped it.[35] Earth then began to cool once again.

———

Rainforests ebbed and flowed across Australia like a tide around slowly emerging islands of sclerophyll forests. The Australian plant family Proteaceae evolved somewhere near the seaway between Australia and Antarctica then crept across the continent, budding off into different species. The old Gondwanan rainforests doggedly tracked the cool and wet conditions of their birthplace by remaining at high latitudes or moving up mountains, such as those of the Wet Tropics and, later, New Guinea and Mount Kinabalu in Borneo.[36] A few rainforest lineages also made cautious advances into the open environments that came to dominate the continent.[37]

Deep within the Wet Tropics, on one of the western flanks of lofty Mount Bartle Frere, lives a cluster of giants, sisters to eucalypts. To find it, you must first locate the narrow track closed in by rainforest and then hike for a few kilometres. The slope leading to the cluster is quite steep. My first visit was with a botanist friend who knew the trees, and I called out to him as I slogged up the slope, looking down at the ground, 'Where are these trees?'

'Right here,' he said rather proudly, leaning on the trunk of a monumental tree that towered over the surrounding forest, its gnarled red bark vaguely reminiscent of a eucalypt. I was awestruck. Only scientifically described in 2002, *Stockwellia quadrifida* has been around for at least the last 40 million years.[38] It was from an ancestor like this that eucalypts made a run for it into the expanding dryer environments, while sister *Stockwellia* continued to track the ancestral pathways halfway up a mountain in the far north of Queensland.

At Tingamarra time, Thiaki was a multistoreyed forest with Proteaceae and emergent Araucariaceae, such as *Wollemia nobilis* (Wollemi pine), and at least 50 flowering genera from the families that we planted in our restoration plots: Cunoniaceae, Elaeocarpaceae (*Elaeocarpus, Sloanea*), Lauraceae (*Cryptocarya, Endiandra, Litsea*), Myrtaceae (*Rhodamnia, Syzygium*) and Proteaceae (*Darlingia, Lomatia, Stenocarpus*).[39] Leaping from trunk to trunk, ferreting around in the litter, and flying through the vine-tangled trees would have been many newly evolved mammals, all (apart from bats) derived from one common ancestor.[40]

Today, there are four groups of Australian marsupials: Dasyuromorphia (insectivorous marsupials, including carnivorous thylacinids); Peramelemorphia (bandicoots and bilbies); Diprotodontia (kangaroos, possums, koalas and wombats); and Notoryctemorphia (marsupial moles). Immediately after the Chicxulub impact event, new species were soaked up into emptied niches as if there were no tomorrow. First, the ancestor of Diprotodontia split with the ancestor of the other three groups. Carnivorous marsupials then went their own way, followed by the bandicoots separating from the marsupial mole:[41] astonishingly, this swimmer of desert dunes started its career as a bandicoot-like creature that burrowed into rainforest litter with sturdy front legs adapted to grab lizards, worms and frogs.

Meanwhile, the ancestor of koalas and wombats split with the ancestor of possums and kangaroos. Kangaroos then separated from possums, coming down from the trees around 50 million years ago in the form of several species of small potoroo-type animals that ran, jumped and climbed low branches using all four legs.

Supercharged in the Eocene hothouse,[42] ancestral songbirds flitted from flower to flower, and ground frogs[43] hopped amongst the skink-slither litter, as did now-extinct terrestrial crocodiles. Early python-like snakes killed with deathly hugs. Thiaki's canopy closed over for the first time as competition among plants led to a race for nutrients and light. In this low light, no wind and, importantly, dinosaur-free environment, plants were able to invest in larger fruits. These, in turn, attracted frugivorous mammals and birds, triggering a coevolutionary spin that continued to change both plants and animals.

Thirty million years would pass before environmental conditions allowed fossils to accumulate once again. And to get a view of this time, palaeontologists had to literally blast a tunnel to the past.

———

BOOM! My son pressed down on the detonator with eyes as wide as only a four-year-old's can be. Twenty metres away, a chunk of hard limestone as big as a small car split neatly in two. It had been intricate work placing the light explosives in such a way that the detonation would split the rock and not blow it to smithereens so it merged with the dusty atmosphere of semi-arid northern

Queensland. It was 1988, and my son and I were visiting the River-sleigh fossil site.

We approached the split rock with anticipation. Embedded, almost magically, on one of the surfaces was a whole skeleton of a huge bird, a dromornithid, or thunderbird. With stubby vestigial wings and massive hind legs bearing strange hoof-like toes, thunderbirds grew up to 3 metres high and weighed half a tonne—or more.

Twenty million years ago, this thunderbird had taken an incautious step on a fragile crust of limestone on the edge of a slippery-sided pool, which collapsed, tipping the bird in. Gathering crocodiles would have waited until the thrashing bird drowned before ripping into the flesh as the carcass sank to the bottom, where its bones, picked clean by turtles, slowly settled into the limy mud to be mineralised as stone when the pools filled with sediment.

Analysis shows that thunderbirds were more closely related to, of all things, chooks than to the other behemoth birds of Australia, the emu and cassowary.[44] *Lithornis*, it seems, had a European sister who had also migrated to Australia, abandoned flight and changed into an even bigger giant with huge powerful legs and tiny vestigial wings. *Lithornis*'s sister had left a trail of behemoths from China to Europe,[45] which are the closest relatives to Australia's thunderbirds. In the mid-Eocene of 40 million years ago, thunderbirds were a common creature across the landscape, with at least seven different species known. The last of their kind became extinct just 26,000 years ago[46]—40,000 years after Aboriginal people arrived in Australia.[47] Thunderbirds are sometimes called Mihirung, a western Victorian Aboriginal name for a gargantuan bird[48] that existed in the Dreamtime.

Twenty million years ago, thunderbirds stalked the forests of Thiaki alongside sprinting crocodiles that ran down their prey on land. Pythons vined among the multilayered canopy laden with epiphytes and dripping with lianas. The forest floor glowed with fungi. Mosses and ferns grew through the rotting litter. Logrunners such as the chowchilla enlivened the mornings and evenings with their cool jazz riffs.

If the fossils of Riversleigh are anything to go by, at night Thiaki would have been filled with chorusing cicadas, chirruping crickets and at least eight different types of croaking frogs. Over 60 different mammals—twice the number of species surviving today—hung out in the trees, including several koala-like species, and at least 24 different species of possums[49] comprising six different ringtail possums, a brushtail possum, a petaurid possum, as well as pygmy possums, feather-tail possums, cuscuses and several now-extinct groups. Stalking the forest floor were medium to small marsupial lions, including micro-lions—essentially carnivorous wombats. There were also thylacines and strange now-extinct mammals with no clear affinity that mirthful palaeontologists have named 'Weirdodonta' and 'Thingodonta'.[50]

It was a dangerous place for the two relatives of the modern musky rat-kangaroo, *Hypsiprymnodon moschatus*, which poked about on four legs looking for insects, fruits and seeds, using their prehensile tails to carry nesting materials. While escaping micro-lions, they would gallop at high speed across the litter-strewn forest floor. Their extraordinary similarity to today's rat-kangaroo across 20 million years is probably due to the enduring rainforest habitat.

Defying the ground were many magnificent types of leaf-nosed bat, ghost bats, free-tailed bats and other micro-bats.[51, 52] And during the day, the forest was filled with sound as the world's songbirds warmed up for a global takeover.[53] A sleek marsupial mole swam through the litter and bothered the soil, which was also rifled by many different types of forest bandicoot and seven different kinds of rat-kangaroos. There were several different types of carnivorous marsupials, such as quolls, devils and small antechinuses. Tellingly, there were also a few large diprotodontid grazers, suggestive of the beginnings of open forest. Many different types of macropods, including a carnivorous one the size of a grey kangaroo, galloped about—hopping was still a thing of the future for the ancestors of Mabi.

These were the glory days of Thiaki and were probably a response to a massive volcanic event[54] at the Columbia River Basalt Group in the Pacific Northwest, associated with the Yellowstone hotspot,[55] which spewed out 175,000 square kilometres of lava and carbon dioxide. At the same time, Earth was being pulled by Jupiter from a more or less circular orbit into an ellipse where, for part of the time, Earth was closer to the Sun—enough to help warm the planet.

The rainforest was expansive, rich and stable, and niches were finely divided and full. Today, this sort of wonderland is only found in the remnant lowland rainforests of Borneo[56] now rapidly being cleared for oil palm.

———

Thiaki's glory days were not to last. They had been an aberration anyway on an Earth barrelling towards an ice age that, ironically, Australia had helped to trigger by tearing away from Antarctica. South America finished the job by ripping a gash in the Antarctic Peninsula, unleashing an icy current that spun around the polar continent, effectively freezing it. In the northern hemisphere, the Tethys Sea—which had shunted warm waters across the northern part of the planet—slammed shut when Africa collided with Eurasia, and India into Asia. The Zagros Mountains emerged, slashing across the Middle East and, with the Himalayas, shedding colossal amounts of sediment that trapped and buried carbon dioxide. This drastically lowered carbon dioxide in the atmosphere.[57] Orbital forcing—the way Earth tilts and wobbles on its axis, and the shape of its orbit around the sun—all helped to tip Earth from greenhouse into icehouse conditions.[58]

Seasonal cool conditions of the Oligocene, around 27 million years ago, had already replaced the hot Tingamarra times, allowing sclerophyll plants such as *Acacia* to spread, and open forests to appear across inland Australia. Some older plant communities were left behind, becoming pinched off and squeezing into habitats resembling the deep past. The Cretaceous-like nook of *Wollemia nobilis* (Wollemi pine) in the rugged Blue Mountains flanking Sydney, and only discovered in 1994, is one of these.

Other communities, particularly in the northern lowlands of Australia, were subject to invasion from Asia. These invaders began to arrive even before Australia split from Antarctica. The genus *Mallotus*, for instance, which is found in Thiaki, dates to around 35 million years ago when Australia was still drawing away from Antarctica, and northern Australia was 450 kilometres away from Asia.[59] Other Asian genera, such as *Acronychia* and *Litsea*—species of which we have planted in our restoration plots—have spawned rich endemic clades in Australia.

A patched environment of woodland, open forest and rainforest was a bonanza for evolution. Wombat-like creatures of all sizes evolved to browse in the open inland forests, nervously watching for the marsupial lions and early thylacines. Several types of possums, including pygmy possums, adapted to a variety of environments and loitered in the trees; a suite of potoroo-type kangaroos pottered about on the ground. Pythons, turtles and crocodiles skulked in swamps, and bats soared from caves and across treetops.

Across the other side of the world, our distant ancestor, who had recently split from gibbon relatives,[60] was hanging around in the trees of Uganda with dozens of other species of apes.[61] No doubt they were all more concerned with the next meal than the effects of the pull of gravity from the gas giants Jupiter and Saturn that were then shaping Earth's orbit and their own evolutionary trajectory.

———

Australia continued its drift towards Southeast Asia. From Australia's north-west edge in the proximity of New Guinea (which still lay under the sea) jutted a long promontory, called the Sula Spur,[62] reaching out towards Asia. It was this promontory that made first physical contact with the margins of Asia around 20 million years ago, pushing up the jewel-like archipelago of Wallacea in eastern Indonesia. For songbirds, this was the evolutionary yellow-brick road.

An ancient mariner sailing to Asia along the Sula Spur from northern Australia would have been baffled by a screen of volcanic micro-continents. It would have been a spectacular tropical archipelago, covered in rich rainforest washed by azure seas, where corals and fish flourished in shallow waters. The precursor of what was to become the largest archipelagic system in the world, stretching across eastern Indonesia to the Pacific Islands, the Sula Spur and the islands of the incipient New Guinea were the testing ground for new species that would later make Wallacea a bewildering theatre of life.

As the Sula Spur collided with fringes of Asia, it threw up the perfectly coned volcanoes of the Banda Islands of Maluku. Chunks of the Lesser Sunda Islands and Sulawesi were riding the rift north as the Sula Spur began to swing and collide with new volcanoes that were pushing up eastern Indonesia. A mishmash of intermingled islands, unmatched in the world, acted as new habitats and were not only drivers of evolution, but also stepping stones from Australia that led to Africa and Asia.

Overhead, the ancient mariner would have observed flocks of birds winging their way from the vast southern continent. The ancestors of most of Thiaki's songsters—and most of Thiaki's birds are songbirds—revelled in the newly created worlds. Perhaps due to their ability to sing and communicate efficiently with each other,[63] the supertramps quickly speciated into new niches, outnumbering other birds so that passerines flooded the world with song.

Back in Australia, the songbirds held their ground against Asian birds flying the other way. The grass birds, however, were an exception. Grass was a recent invader, becoming a significant part of Australia's vegetation about five million years ago.[64] This was good timing for the grassland estrildid finches that evolved in India as the

Himalayas soared upwards and created an avalanche of new species, which spread west to Africa and east to Australia.[65] India had long before docked against Asia in Tingamarra times, but the two land-masses merely snuggled. The union became serious only about 14 million years ago, leading to the birth of the high Himalayas and a suite of new species—such as the ancestor of Australia's finches—about 11 million years ago. The ancestors of the red-browed finch and chestnut-breasted mannikin arrived in Australia around nine million years ago[66]—flying over the galaxy of archipelagos to get here and finding their way ultimately to Australia's spreading grasslands that today lap Thiaki's flanks.

Flying with them were kingfishers. It seems that they, too, evolved during the Himalayan orogeny around 16 million years ago,[67] flying over the Sula Spur to Australia all the while sprinkling species over the islands of Wallacea, the incipient New Guinea and Australia, where they morphed into a suite of characteristically Australian kingfishers such as the forest kingfisher, sacred kingfisher, laughing kookaburra and blue-winged kookaburra. The ancient mariner might have been surprised to see varanids swimming from island to island. They ultimately made their way to Australia sometime after 20 million years ago,[68] evolving into the largest and deadliest lizards in the world, Megalania. These would return to Wallacea as the Komodo dragon.

———

As Australia drifted north, birthing fresh and gleaming island chains, plants continued to percolate both ways, mostly carried in the slow but steady drip of droppings from peripatetic birds. In the beginning, this direct seeding had little impact on the richly patterned landscapes of rainforest with its full niches. But around the beginning of the Pleistocene, around two million years ago, Australia's rainforests began to shrink in response to climatic pressure that was not felt by the more stable and expansive rainforests of Indonesia.

Asian plants began to invade Australia's weakened rainforest niches, particularly during interglacial periods when rainforests were attempting to bounce back from refugial pools. *Toona ciliata* (red cedar) arrived during a recent ice age, perhaps blown in by the strong monsoonal winds that were intensified by the massive ice sheets squatting over much of the northern hemisphere.

So successful were these Asian ancestors that they now comprise nearly half of lowland rainforest lineages in the Wet Tropics, from where they weaved rich endemic lines.[69] In the monsoon tropics of Cape York, Northern Territory and Western Australia, with a topography and climate like those of Southeast Asia, Asian rainforest elements came to dominate. Further south, fewer Asian immigrants were able to take hold, but those that did now have high levels of endemicity. Those attempting survival further south found it tough in regions of unfamiliar topography and climate.[70] By the time they reached Tasmania, the chances of survival were zero.

———

As the Himalayas began their final leap, they caused massive cooling and aridity in Asia, and the Arctic began to freeze over. These changes reverberated around the world. In Australia, they resulted in more seasonal and drier weather patterns. A steep temperature difference between the poles and the equator whipped up strong winds that sent the ocean churning and scooping up deep nutrient-rich waters, driving algal blooms that gobbled carbon dioxide from the atmosphere. Grasslands—which absorbed less heat than forests—spread across the northern hemisphere. By seven million years ago, Earth was cooling[71] fast.

At the same time, Australia's northern bumper began smashing into the edge of Asia, smearing onto it a collage of crustal fragments, and finally pushing up the New Guinea highlands five million years ago. The deep-sea passage between Indonesia and Australia was strangled, blocking the flow between the Indian and Pacific Oceans that was helping to maintain equable global climates. Currents were instead shunted southward along Australia's eastern margin, down past mid-latitude Tasmania and onward towards Antarctica.[72] Like a whip to a flywheel, this East Australian Current drove more power into the bitingly cold Antarctic Circumpolar Current that was engulfing Antarctica. Globally, temperatures and rainfall continued to plummet.

In Australia, families of rainforest animals unknown to us today—Yaralidae, Yalkaparidontidae, Pilkipildridae and Balbaridae (distantly related to bandicoots, possums and kangaroos)—withered and died as rainforests began to shrink from the continent's interior. This proved a bonanza for the ancestors of kangaroos and wallabies, including Mabi; around 13 million years ago, they bounded into the new sunny forests and woodlands, along the way morphing into new species. The rainforests were left to *Hypsiprymnodon* rat-kangaroos and potoroos. Over time, kangaroos adopted an efficient hopping gait and developed a specialised stomach to digest the new tougher vegetation and grass.[73]

One small pademelon-like wallaby preferred to graze closer to home around the edges of rainforests, exploring rocky and scrubby environments where they darted in low crouches with extraordinary speed and agility, even climbing small leaning trees.[74] Around ten million years ago, this species went back to the trees as modified kangaroos with a thickened muscular neck built for climbing (and falling), teeth meant for shearing leaves rather than grinding grass, a pendulous long tail for balance, strong front legs and broader hind feet, all with curved claws and spongy soles for gripping.[75] The yellow-footed rock-wallaby and the little nimble nabarlek of rocky Arnhem Land in the Northern Territory and the Kimberley region of Western Australia are the tree-kangaroo's closest ancestors.

———

For a brief interval in the Pliocene, around five to three million years ago, Earth warmed[76] once again when ocean gateways cracked open between North and South America, allowing the transport of tropical water around the globe. Atmospheric carbon dioxide was at 400 parts per million, and the average temperature was 2–4 degrees Celsius warmer—more or less the same as predicted for later this century.[77] But there was a big difference. In the Pliocene, wholesome and unfragmented landscapes triggered not extinction, but a radiation of species—including a battalion of bandicoots, from rainforest species to relatives of the desert bilby.

Thiaki animals were holding their own.[78] Frogs abounded in numbers like the glory days. Boyd's forest dragon, a lizard that changes colour to blend into the rainforest, and other lizards were plentiful. Carnivorous marsupials thrived, including thylacines, Tasmanian devils and at least two species of marsupial lion; phascogales and smaller carnivorous marsupials kept the abundant bandicoots on their toes. Cuscuses and at least 14 species of possums—brushtail, ringtail, feathertail glider, striped and pygmy—dangled from trees, including the giant *Pseudokoala*, a possum not a koala, though as big as one. At least six types of placental murid rodents had also arrived from Asia by this time. Further inland were several species of megafauna including a giant wombat, a diprotodont and a couple of giant kangaroos.

After returning to the trees, Mabi's close relatives evolved into many different types of tree-kangaroos. These were found across Australia—the recent modern distribution is a poor mirror of their past range[79]—including in what is now one of the most arid environments in the world, south-east of Birdsville in northern South Australia. Here, a giant tree-kangaroo called *Bohra*, the size of a small brown bear, vied for branch space. After their split from rock-wallabies, Australian tree-kangaroos hopped across to New Guinea, and evolved with the rising Central Ranges, Huon Peninsula and Bird's Head Peninsula into 11 species.[80]

Across the other side of the planet in the wide Ethiopian plains, the common ancestor of humans and chimpanzees had also appeared.

While the exact identity is unknown, a close relative, *Ardipithecus ramidus*, known from a nearly complete skeleton, provides a snapshot of its character. *Ardipithecus* preferred woodlands and open forests, where it climbed slowly and deliberately—more clamber than climb—and walked with a slightly stooped posture. *Ardipithecus* ate broadly, and its brain was larger than its ape predecessor, possibly to allow for more intensive and careful search-oriented feeding accompanied perhaps by more social food transport and sharing behaviours.[81]

————

The brief warm interlude was shut down by the closure of two oceanic gateways: the Isthmus of Panama and the Indonesian Throughflow.[82] Like a one-two punch, the shutting of these two oceanic gateways resulted, first, in the North Atlantic heating up at the expense of the southern hemisphere in a case of heat piracy, resulting in increasing Antarctic glaciation. The tectonic reorganisation in Indonesia blocked the flow from warm and salty South Pacific water to fresher and cooler North Pacific subsurface water. This triggered the northern hemisphere glaciation. The uppercut came when the highlands of New Guinea soared skywards to altitudes of nearly 5000 metres above sea level, sending a dry rain shadow across northern Australia.

The conditions for the Pleistocene ice age—with its extinctions of giant birds, reptiles and mammals—were set. All it took was the cosmic flutter of a butterfly's wings—the way Earth wobbles on its axis—to tip the planet between glacial and interglacial conditions like a regular 41,000-year heartbeat. Over the Pleistocene, this dizzying climatic yoyo resulted in regular glacial periods punctuated by 42 short interglacial periods.[83] The current epoch, the Holocene, is an interglacial continuation of this yoyo. During these cycles, great continental ice sheets waxed and waned, in turn locking up and releasing water. Sea levels rose and fell to as low as 125 metres below the present level.[84]

In the northern hemisphere, ice sheets up to a kilometre thick accumulated over much of North America and Eurasia. There were no major ice sheets in Australia. Instead, during glacial periods, the climate became dry, cold and windy; during interglacial periods, it was warmer and wetter. The vegetation likewise pulsed, like a heartbeat, with open shrublands and grasslands expanding during glacial periods, and forests and woodlands expanding during interglacial periods.

It was finally in these extreme conditions that Ernie's country-man, Lumholtz's tree-kangaroo, *Dendrolagus lumholtzi*, and its sister, Bennett's tree-kangaroo, *Dendrolagus bennettianus*, sprang to life. Once widespread, tree-kangaroos vanished across the broad expanse of Australia. The only survivors were those clinging to the rainforests of the Wet Tropics, which had shrunk to an archipelago of isolated green pools as the Pleistocene dawned. Today, only one tiny rat-kangaroo is witness to the early goings-on of tree-kangaroos.

# CHAPTER EIGHT

# Nothing is nothing

While Thiaki was being pummelled by the Pleistocene climate shifts, half a world away in Africa the Pleistocene spawned a change of another kind. An auspicious, long-legged, ground-dwelling, upright ape emerged. Normally an African animal would not warrant Thiaki's attention. But this one was worth watching. An enlarged brain allowed the ape to hold a stone with dextrous fingers and see the tools that could be made from it to help feed the ape's high-energy lifestyle. *Homo erectus* families had tamed fire and probably sat around warming hearths, perhaps communicating through speech.[1] Shortly after appearing in Africa two million years ago, *Homo erectus* left for a more cosmopolitan life and travelled across Asia. After traversing high mountain passes, deserts and woodlands, the ape finally reached a sparkling new archipelago in equatorial seas some 15,000 kilometres away from its birthplace.

Concepts of Africa and Asia were, of course, meaningless to this ape. A savanna specialist, the early human instead followed 'Savannahstan',[2] which spread from west Africa through the fertile

Indus Valley and the richly wooded foothills of the Himalayas to north China and down into Southeast Asia.[3] Here, clans of *Homo erectus* walked to the edge of a vastly expanded ice-age Asia, where a forested corridor had replaced the sea that had drained away to expose Java and Borneo as part of the mainland. The Pacific Ocean churned south through a narrow seaway between Sulawesi and Bali, gushing into the Indian Ocean.

Java itself, rising as a result of the ongoing collision between Australia and Eurasia, was a series of volcanoes surrounded by coastal swamps in which wallowed tiny hippopotamuses, mastodons, tortoises and deer. As the island continued to rise, the swamps were replaced by chains of volcanoes and alluvial plains that were covered by open woodland and gallery rainforest. By a million years ago, the scene was more akin to *Out of Africa* than tropical Asia, with tigers, leopards, wild dogs, hyenas and *Homo erectus* preying on herds of *Stegodon*, hippopotamuses, pigs, deer, cattle and antelopes.[4]

The crisp high peaks of archipelagic Indonesia silhouetted on the horizon must have seemed so close across the sea that the new human could imagine reaching out to touch them, particularly during the four exceptionally massive glaciations around 650,000, 450,000, 140,000 and 20,000 years ago when the ocean plunged, according to some estimates, 140 metres below the present level.[5] Drawn by the allure of new hunting grounds, some intrepid travellers may have tried to reach them by cobbling together logs into a makeshift raft. Or perhaps groups of humans clinging to rafts of debris were simply swept out to sea by tsunamis generated by the frequent earthquakes and volcanic eruptions of the region.

In a place where the world's biggest ocean was attempting to squeeze through a narrow channel, animals rarely survived the churn to make landfall. Those few that did, however, landed on strange new worlds with weird endemic island faunas of miniature elephants and giant lizards. Here, like in other places where animals were subjected to the pressures of island life, very early humans had also dwarfed over aeons into new species, such as the diminutive

*Homo floresiensis* of Flores and *Homo luzonensis* of the Philippines. Did they also make it to Australia?

In the meantime, back in Africa another human species with a penchant for travel emerged around 700,000 years ago.[6] These Neandersovans[7] split into Neanderthals, who lived mainly in Europe, and Denisovans, who travelled to Asia, where they left their remains in Denisova Cave in the Altai Mountains of Siberia (a find only announced in 2010[8]). The two were known to interbreed, and they produced viable offspring.[9] A promiscuous lot, it seems Neandersovans also interbred with the old humans, *Homo erectus,* or their variants, who had left Africa over a million years before.[10] The two populations are more distantly related than any pair of human populations known to interbreed, raising the question of how many species represent humans.[11]

———

Back in Australia, a new suite of animals was emerging into the more openly forested plains and woodlands. Large tiger-sized thylacines stalked half-tonne wombats; kangaroo species flourished, some evolving into giants that grazed the first grasslands; forest wallabies and small, modern, carnivorous marsupials appeared; three species of koala evolved to munch on the tougher leaves; a gigantic wombat, *Diprotodon optatum*—the largest-ever marsupial, weighing in at 3 tonnes, which is bigger than a hippopotamus—trundled about; forest bandicoots and bilbies evolved; and the only grazing bandicoot (the others are omnivores), the pig-footed bandicoot, emerged.

But as the severity of the glaciations increased, bare dust and sand were stripped from inland Australia and piled into an anticlockwise swirl of longitudinal sand dunes approximating the wind direction of the subtropical anticyclone systems that were sweeping across the continent. Along the coast, the middle Pleistocene rainforests of 780,000–120,000 years ago still thrived, but they were reduced to coastally aligned archipelagos—albeit still more extensive than they are today—and separated by open forests and grassy woodlands.

At Mount Etna, 1000 kilometres south of the Atherton Tablelands, extensive fossil deposits provide a rare insight into this time.[12] The reduced rainforests were still harbouring an extraordinary suite of animals, including megafauna such as *Thylacoleo*, a leopard-like tree-dweller; 500-kilogram *Palorchestes*, which dragged down vegetation using its long claws and trunk; and *Diprotodon*, which may have stomped through the rainforest in a similar fashion to forest elephants. There were also forest dragons, frogs, a pademelon wallaby, rainforest-loving white-tailed rats, over 14 different possums and at least four different tree-kangaroo species.

But then a particularly severe drying glacial cycle blasted the area, and all the megafauna—together with 80 per cent of the other mammal species, the frogs and rainforest lizards—disappeared from central Queensland. All the megafauna became extinct, but 11 of the possums and all of the tree-kangaroos found refuge in New Guinea, which has mostly been the northern rampart of Australia, harbouring an extraordinary range of climate zones and habitats from savanna and extensive rainforests to alpine zones. While the green ringtail possum, *Pseudochirops archeri*, claimed Thiaki, four or five of its close relatives migrated north across vegetated plains to New Guinea.

Extinction and evolution normally go hand in hand: when the rainforests of the mid-Pleistocene were in full bloom, for instance, perhaps 12 per cent of animal species became extinct and were replaced by about the same number of new species. But Pleistocene climates on Earth were shaky at best. When aridity really bit, it dried up the number of available niches so that only about half of the 65 per cent of species that became extinct at Mount Etna were replaced.[13] Instead of rainforest animals, the area was occupied by the desert-adapted pig-footed bandicoot, bilby, desert bandicoot, carnivorous marsupials, hopping mice and desert dragons.

———

In Africa, the Pleistocene ice age continued to fuel the hominin maelstrom, which produced the ultimate hominin supertramp—contemporary humans—250,000 years ago. Those who travelled into Europe interbred with Neanderthals, possibly even assimilating them into the human population of that region.[14] Clearly there were no strong biological or behavioural barriers to sex with other types of humans, and it seems groups of different humans bred whenever they came into contact.

Not everyone went to Europe, though; some instead turned to the rising sun and filtered through the forests of Asia—and onwards to Australia. By the time they got to Southeast Asia, perhaps 70,000 years ago,[15] *Homo erectus* had been extinct for at least 40,000 years. This heroic two-million-year-old human species had managed to cling to the edge of Asia alongside dying remnants of an archaic savanna-loving fauna. *Homo erectus* finally succumbed to the rainforest flora and fauna of the penultimate warm wet interglacial period[16] of 130,000 to 110,000 years ago, along with Lydekker's deer, *Axis lydekkeri*; a prehistoric water buffalo, *Bubalus palaeokerabau*; an elephant, *Elephas hysudrindicus*; and a relative of elephants, *Stegodon trigonocephalus*.[17]

Even so, when contemporary humans got to Southeast Asia, they did not find a *terra nullius*: another modern human, Asian Denisovans, had got there earlier and had been there for millennia, long before the extinction of *Homo erectus*. It is perhaps not unexpected that contemporary humans of Asian descent would mix with Altai Denisovans, the two being more or less in the same place. But the fact that New Guinea—8000 kilometres away and across an archipelago larger than Europe—was the main red-light district for Denisovan interbreeding was an unexpected and surprising reveal of modern Papuan genomes.[18] And, astonishingly, this interbreeding was not from one Denisovan group but from two quite different Denisovans—separated by thousands of years.

It is a monumental genomic 'whodunnit' or, more correctly,

'who was doing it'. Shortly after Neanderthals and Denisovans split 400,000 years ago, it seems that Denisovans themselves split into at least two groups: one at 363,000 years ago, and another at 283,000 years ago. These two deeply divergent and geographically isolated peoples apparently never met. But each interbred with the new wave of modern humans.

Clearly adaptable, Denisovans spread across Asia and into Southeast Asia, traversing continental steppes, slogging through mountain passes and apparently rafting tropical equatorial islands, even crossing the robust water barriers of Wallacea. Here, it appears, they met Papuans. The first tryst is estimated to have happened somewhere in the islands of Wallacea, or on New Guinea itself, as early as 60,000 years ago. And the last as recently as 14,000 years ago. There's also another echo in the Papuan genes, a residual smudge of unexplained but very archaic DNA. Could it be the last word of *Homo erectus*?

In this astounding picture of human evolution in Southeast Asia, can we even be sure that contemporary humans were the first to reach Australia?

———

Whatever the case, when our direct ancestors reached the edge of Asia at least 70,000 years ago,[19] the sea level was about 80 metres lower than now,[20] exposing 17 potential crossing points. These were places where intrepid mariners could in most cases see the next island, and from there, the next, and so on all the way to Australia. Between the edge of Asia at Borneo, for instance, you could see Sulawesi, and from there you could see the sparkling Maluku Islands—first Obi, then Buru, Halmahera and Seram. From here the towering Bird's Head Peninsula of New Guinea would become visible. Or you could launch, instead, from Java. Bali is just a couple of kilometres away. Travelling east along the chain of linked volcanic islands would be easy enough: through Flores, where the tiny

*Homo floresiensis* resided, and then taking to the sea across narrow straits separating Alor and Timor.

From Timor, the ancient mariners would carefully read the currents and wind, and drift to the chain of habitable resource-rich islands that had emerged from the sea 300 kilometres off the northwest coast of Australia; when seen from the high points of Timor and Roti, these islands wavered like irresistible mirages on the horizon. Large numbers of birds flew overhead to and from the area, and occasionally there would be a waft of smoke from lightning-sparked bushfires.[21] If you paddled, it might take two to four days to get there. Anyone daunted by the 90-kilometre journey across what were the treacherous currents of the Indonesian Throughflow, funnelling a significant volume of the Pacific Ocean between Timor and Australia, could continue instead from Timor to the Tanimbar Islands and then raft to the wooded plain that joined New Guinea and Australia, roughly in the area of the Aru Islands today.

One thing is certain: settlement of Australia 65,000[22] years ago was not accidental.[23]

Genetic modelling suggests that Australia needed to be colonised by a minimum of 1300 to 1500 people, similar to the minimal numbers estimated to make a viable population in New Zealand or the Americas. Any fewer people would expose the population to random catastrophic events or to inbreeding—both ultimately leading to extinction. Did the first people come all at once? And under what circumstances: pure wanderlust; pushed by an occupying force of Denisovans; an initiation dare; or simply because they could see Australia? Or did they come in smaller groups over a 'survival window' of several generations?[24]

Genomic signatures suggest that there was one main colonising event that occurred over a narrow window of seven hundred years or so. Within this event, two quite separate arrivals are indicated by the different genomes of New Guineans and Aboriginal Australians:[25] one group, the New Guineans, likely came via a northern route through Sulawesi and the Maluku Islands to New Guinea, and the other group, the Aboriginal Australians, via a southern route along the chain of islands leading to Timor.

People then cautiously filtered across the landscape, favouring life-supporting environments, and building knowledge, culture and technology.[26] Over millennia, this careful and respectful occupation may have appeared as purposeful colonisation. People could have diffused throughout the continent in less than 5000 years and reached populations of perhaps six million people, given the richness predicted in the Late Pleistocene environment before the withering last glacial period.[27] Modern human populations entering the Alligator Rivers region of the Top End 65,000 years ago, for instance, would have experienced a relatively dry, but not arid, environment with rich pockets of monsoon vine-forest vegetation.[28] Inland, perennial rivers fed mega-lakes—some of the biggest in the world—that were teeming with wildlife.[29]

In these relatively equable conditions, people not only filtered down the coastlines of eastern and western Australia in parallel clockwise and anticlockwise movements[30] but also followed super-highways of resource richness including through the interior of Australia.[31] Three different groups, who had been separated for perhaps 10,000 years, appear to have met in South Australia 50,000 years ago in an archaeological instant and in what must have been a mind-boggling moment. Meanwhile, eddies of people peeled off down the Murray–Darling system and into the central and western deserts, following persistent water supplies of springs, swamps, seeps or cryptic gnamma holes—small rocky cisterns—with precarious and precious rainwater.

After diffusing across Australia, strong and deep splits in the genetic structure confirm that people stayed put, oozing into their respective landscapes: the messages encrypted in the Aboriginal Australian genome indicate an extraordinary 50,000-year geographic fidelity. This deep forever relationship with landscape, culturally expressed through attachment to Country and reinforced through Dreamtime narratives, is intimate—almost cellular—allowing people to survive embedded in their sentient landscapes, through the extremes of climate change.

———

It was not a place for the faint-hearted. A rare find of fossils in the headwaters of the Fitzroy River[32]—a Great Barrier Reef catchment area 600 kilometres to the south of Thiaki—reveals a sharp-toothed fauna dominated by marsupial lions that hunted five or more species of kangaroos, including a giant species that weighed around 300 kilograms and thrived on the growing grasslands. Saltwater crocodiles were accompanied by even larger freshwater genera, *Pallimnarchus* and *Quinkana*. Giant monitor lizards—including the 600-kilogram, 7-metre-long *Megalania*—stalked the woodlands, dripping saliva as they ran down laggards in herds of grazing horse-sized palorchestids, and diprotodontids, thriving on extensive grasslands.

The climate was unpredictable, with periods of wetter monsoons and long periods of drought when the monsoons failed. Like a fractious child before settling, it threw different combinations of hot, cold, wet and dry. Ecosystem response to this complexity was itself unpredictable. At 50,000 years ago, complex rainforests, like today's, sent their pollen drifting into the swollen rivers of a brief wet period, and Thiaki's mountain rainforests expanded with the cooler temperatures of the looming glacial period.[33]

But then about 48,000 years ago, there was an abrupt and sustained drop in effective rainfall punctuated by diabolical droughts associated with El Niño events.[34] Megafaunal extinctions in the Great Barrier Reef catchment areas coincided with this deterioration.[35] The effects were regionally staggered in time and space: Lake Eyre Basin felt the impact first, its great mega-lakes shrinking to salt, then the Murray–Darling Basin and finally the Fitzroy Basin. The saltwater crocodile survived in viable coastal habitats, while the freshwater *Pallimnarchus* and *Quinkana* disappeared with their habitats. By about 40,000 years ago, dry rainforest plants and sclerophyll forest had largely replaced the complex rainforest, which shrank into valleys, cowered against mountains or cringed close to the coast. Grasslands shrivelled under sustained El Niño drought, starving the dependent herbivorous megafauna. By the time grasslands were able to make a brief comeback 35,000 years ago, the megafauna had vanished.

—

The startling tectonic timeframe of the oral history of Aboriginal people still whispers of this time. It is arguably most obvious in the Kimberley region of north-west Australia where archaeological evidence suggests that people first made landfall 65,000 years ago.[36] Well-respected Aboriginal Elder[37] David Mowaljarlai wrote, in the evocative book *Yorro Yorro*,[38] of the Wandjina whose paintings marked an old distant coastline from long ago, and of sacred sites that now lie beneath the Indian Ocean. He spoke of an older painting style from the ice age 'before the Flood', when the land was all one big continent. The distance from the Kimberley to the ice-age coast was then up to 300 kilometres. The climate was arid and windy, and the monsoon—with its promise of regular seasonal rain—had all but ceased.

The population was sparse; people in small clans clustered around rich refuge areas such as the Kimberley,[39] where they executed stunningly elegant paintings in the Gwion style. A minimum date for these is around 17,500 years,[40] but they could be much older— from a time when megafauna roamed Australia. One painting,[41] in fact, shows a hunter engaged with a large striped animal that has massive forequarters. The hunter has a two-handed grip on a barbed spear that is buckling with the force of the fight, suggestive of either spearing or fending off an animal much bigger than himself—perhaps a marsupial lion, *Thylacoleo carnifex*. The largest Australian carnivore, it might have weighed 160 kilograms with cheek teeth that formed long shearing blades.[42] Thought to have become extinct perhaps 46,000 years ago, it may have hung on in ice-age refuges where, with the disappearance of its natural prey, it turned to humans.[43]

About 17,500 years ago, Earth's ice began to melt. It happened quickly, causing sea levels to rise and, in Australia, engulf an area of land the size of Queensland. Easily perceived as a 'flood', the expanded Kimberley was inundated by as much as 130 metres per year or 6 kilometres in a person's lifetime,[44] drowning entire Aboriginal countries. Sea levels stabilised around 8000 years ago, ushering in a period known as Australia's thermal maximum, the warmest period in the last 30,000 years.[45] The life-giving re-establishment of the monsoon and westerly winds replenished sacred Country. The minimum age for Wandjina is 5000 years.[46] According to David Mowaljarlai, the Wandjina waded through flood waters to start life anew.[47] Today, the Wandjina's rumbling thunder heralds the first storms of the monsoons and calls to people with the promise of fresh life.

———

In the north-east of Australia, the first people to view the region of the Great Barrier Reef from the adjacent coastal ranges near where Cairns is today would have seen an undulating expanse, not of sea, but of plains and valleys filled with a patchwork of grasslands, forest and rainforest. These were interspersed with massive freshwater lagoons ringed in by limestone hills, which were coral mounds during previous interglacial periods. By 58,000 years ago, the Great Barrier Reef was high and dry. Coastal rivers ran out across the continental shelf, 40 kilometres east of the present coast, dumping their loads of sediment, pollen and charcoal into the sea.

Today, a deep-sea core drilled into this continental slope seaward of the Great Barrier Reef plucks out a plug of vegetation signatures[48] that are like a series of stills from an old moving picture: jumpy and scratchy but still able to be spliced together to get a sense of the story.

The core shows that the climate at the time of the arrival of Aboriginal people was drier and cooler than today. This is indicated by the coniferous araucarian-dominated rainforest patches and extensive grasslands that thrived with regular seasonal rainfall: grasslands can take the heat, but they wither with no seasonal rain.[49]

In the Wet Tropics, creation stories from coastal rainforest peoples in the Cairns area recall that the sea was at the outer edge of the Great Barrier Reef. Twenty thousand years ago, the land's edge was marked by a steep cliff plunging down into the sea beyond the continental shelf. An ancestral hero called Ngunya lived there. When his son-in-law contemptuously killed too many fish, Ngunya was insulted and caused fresh water to bubble and flow, allowing fish to escape. He didn't stop the water, however, and it covered all of the wooded lands in a tidal wave. While fleeing, Ngunya came across groups of people from different tribes driven together by the flood.[50]

Other stories recall Fitzroy Island, now separated from the mainland by a 30-metre-deep ocean trench—which once formed a mainland promontory—and Green Island, now a small smear on the sea, being vastly bigger. The rainforest peoples also recall a time when an ancestral hero strode across the landscape, naming mountains, waterfalls, people and animals. Occasionally dissolving into the earth and bursting out near waterfalls, he straddled valleys and mountain peaks that are now holiday islands tens of kilometres offshore.[51]

Across the continent, people were thin on the ground in the last glacial period, with some estimates suggesting that there might have been one person per 700 square kilometres[52] (slightly smaller than the size of Australia's capital city, Canberra). They were concentrated in well-watered ranges and near rivers across nine or ten main refugia during the growing climatic instability associated with the Last Glacial Maximum.[53] There might have been as few as 2000 people in the greater region surrounding the Wet Tropics that included a large part of Cape York Peninsula and the Einasleigh Uplands to the west, which harvests the headwaters of the rivers running into the Gulf of Carpentaria.

Their diet was anchored in plants. Pleistocene people must have come to Sahul with an adaptable broad-spectrum toolkit and a deep knowledge of plant use that was a fundamental and culturally transmitted aspect of human survival. The recognition of plants would have been a survival lifeline when they arrived in Australia. The very first meal was likely a preparation of roots and seeds with a side dish of yams and the apex and pith of palms,[54] requiring multistep and labour-intensive processing.

In Arnhem Land, along with the earliest records of people in Australia, are remnants of the first meals.[55] These included bush plums and the fruits of *Buchanania* and *Terminalia* that, to this day, are a favoured food plucked and eaten straight from the tree. Remnants of *Pandanus* show that people were also extracting the rich kernels from the fibrous and mechanically resistant fruits using stones and pebbles. Nuts of Australian *Canarium* species—related to *Canarium indicum*, the widely harvested Melanesian tree crop known as galip nut—were also eaten. Today, you can buy them roasted, peeled or natural from shops and supermarkets.[56] Pleistocene humans seemed to prefer them roasted after extracting them with a single blow from a hammerstone. Different types of *Terminalia*—Kakadu plum, native peach— were also eaten. Today, we know that they are a vitamin C superfood.[57]

———

Contemporary humans were uncannily adaptable. At the same time as the ancestors were wandering the plains east of Cairns and the deep-valley rainforests around Thiaki, their cousins back at the edge of Asia were observing an evening rush of bats streaming in their hundreds of thousands from a vast interconnecting cave complex spread over tens of hectares in the rainforests of Borneo.

Here, at the Great Cave of Niah, the quiet whoosh of bats was an undercurrent to the thousands of squeaking, screaming, chirping swiftlets—all fluttering small wings—waiting for the stream of bats to leave so they could take their place inside the deep caverns, and escape the raptors who dived in and out of the flocks, casually picking off birds and bats.

Inside the massive cavern, people were bringing in plants as well as wood for fires to cook the meat that had been hunted that day. They entered a soaring cavern and stepped onto a platform overlooking a river valley 30 metres below. The entrance of the cave arched up 70 metres, with huge stalactite towers, 3 metres thick, dropping from the soaring ceilings. Trees, climbers and hanging vegetation shaded the cave. The cave would not have failed to impress the first people who found it during a warm period over 50,000 years ago. Since the cave was surrounded by dark moist forests, the people must have entered it tentatively at first. But the cavern was cool and dry, and it soon echoed with chattering voices, laughter, gossip, squeals of children and the sound of digging sticks shovelling dirt to make pits for cooking. The cave floor became a 40,000-year palimpsest of activities from people who occupied it regularly but sporadically when gluts of fruits and nuts—and hence animals—were in the vicinity.

Archaeological digs in the Great Cave of Niah record an extraordinary environmental transition: from warm rainforest, when Sarawak was 15 kilometres offshore in the South China Sea, to cool, wet, coniferous forest and dry savanna when Borneo was part of a vast Asian continent. Then flooding occurred once again, with mangroves growing nearby as the sea reached almost to the cave. The accumulated detritus of millennia shows that there was a focus on catching bearded pigs using deliberate and targeted strategies involving snares or traps.[58] People were, however, broad-spectrum foragers, hunting at least 40 different types of mammals and reptiles, both arboreal (hawks, hornbills, orangutan and leaf monkeys) and terrestrial (Asian brown tortoise, Sumatran rhinoceros, deer, pangolins and cattle), many types of birds as well as freshwater fish and freshwater and saltwater molluscs.

In both Arnhem Land and Borneo, people were preparing the same type of plants, such as palms, yams and *Canarium*. In Borneo, there is also evidence that people were already processing abundant resources of toxic nuts, such as *Pangium edule*, which contains hydrogen cyanide—made edible by fermentation. Today, these kernels make a thick black gravy that is popular across Indonesia in rice dishes, and fish and meat stews.

Unlike *Homo erectus*, contemporary humans found rainforests no barrier to living. On the contrary, people were using a battery of technologies: projectiles such as bows and arrows, blowpipes and stingray spines modified into projective heads with resin to hunt mammals and fish; and techniques such as tuber harvesting and plant detoxification, which required time and complex social inter-actions. People were also burning small patches within and along the edges of the rainforest to attract grazing animals.[59] Because rainforest resources are distributed broadly in low densities, people were highly mobile. With complex socio-economic dynamics, extraordinary levels of ingenuity and forward planning, they had a superlative— certainly holistic—knowledge of the entire landscape. Based on an intimate knowledge of life cycles and territories, people used hunting and gathering strategies that were self-evidently sustainable.

In Sri Lanka, too, the same adaptive flexibility and broad toolkit were applied and moulded for the specialised hunting of arboreal monkeys and giant squirrels.[60] These two animals are now consid-ered the most vulnerable to overhunting—but ancestral Sri Lankans had been hunting them for 40,000 years as a main element of their diet without significant changes to monkey or squirrel popula-tions.[61] Today, animals of the rainforests of Southeast Asia are at a greater risk of extinction than ever before due to unsustainable clearing and overhunting that creates a perverse grassland ecosystem in which not only rainforest species, but also grassland species, are notably absent.

In New Guinea, people arrived and quickly dispersed over a topographically complex region, scaling rainforest-clad, landslide-prone, V-shaped valleys to discover equable montane valleys, such as the Ivane Valley that has archaeological records going back nearly 50,000 years. To get there required a walk from the east up the Kokoda Valley and then over a 4000-metre-high range. With temperatures cooler than today, this meant trudging through freezing subalpine meadows that were home to the *Protemnodon* and diprotodontid subalpine browsers as well as their predator, the thylacine (*Thylacinus cynocephalus*).

Residue on tools shows that the small bands of people were harvesting *Pandanus*, a high-fat and protein-rich food—still an important staple in the Ivane Valley today—which may be one of the reasons the ancient people visited the valley in the first place.[62] These people also ate the yam, *Dioscorea alata*. Since it was found naturally at lower altitudes, people must have transported it up.[63]

———

In Australia, people became enmeshed within ecosystems. The long history of coevolutionary communication between cultural practice and particular ecologics drove evolutionary feedbacks that spurred cultural evolutionary change—so land-use practices became highly finetuned to local environments. Dreamtime and the Law stemming from it are the institutionalisation of such histories of ecological practice.[64]

It's hard to overemphasise the depth of knowledge that people had of the Australian bush. Moment by moment, scratches on trees, a fallen leaf, a broken branch or overturned rainforest litter all speak and tell a story. A human footprint is immediately known. The way the wind whispers in the leaves tells of the type of tree and the season. The turning colour of leaves and flight of birds are signs that yams may be ready to dig up, fruits ready to pick or fish running in the river. For a people who have Country running through their veins, the planting and tending of crops in a place of such abundance would surely be absurd.[65] There is almost nothing that is not put to some use. Nothing is nothing.[66]

I go to my favourite cookbook, Stephanie Alexander's comfortingly thick *Cook's Companion*,[67] which has around 1000 recipes ordered according to ingredients. My book is the first edition, from 1996, so it is a bit wonky at the spine because of constant use. I count the number of vegetables used in the voluminous book— it's around 120 and includes nuts, herbs, mushrooms and salt. I then go to the *Dyirbal Thesaurus and Dictionary Across Ten Dialects*[68] and count the number of times a plant is noted as being edible. It is around 150. This is almost certainly a significant underestimate given that this is not a cookbook, the focus of the list was not edibility, and the informants were several generations removed from the pre-European rainforest peoples.

*The Cook's Companion* includes about 25 different types of animals, such as pig, deer, cattle, duck, kangaroo, rabbit, chicken, pigeon, fish, crab, octopus, scallop and lobster. The animals eaten by rainforest peoples can be classified by habitat: forest floor, forest canopy and freshwater animals found in the forest.[69] European observers noted that about 15 different animals from the forest floor were eaten, the most important being fowl, turkey and cassowary. Aboriginal people freely climbed into the canopy using a length of the abundant *Calamus* lawyer vine about 5 metres long. The vine was knotted at one end and held with one hand, then thrown around the tree to be caught and wound around the other hand. With this secure hold, a foot was planted against the tree, the body was held as far back as possible and the vine loop was thrown up the tree; with momentum, people walked up the tree as easily as a sailor uses a ladder.[70] In the canopy, at least 20 different types of animals were found, including possums, tree-kangaroos, white-tailed rats and pythons. In the streams and swamps were at least eight different types of fish, eels and crayfish, as well as water monitors, geese and ducks.

But perhaps the most remarkable thing about the rainforest diet, though, is that the dominant staple food plants are toxic.

———

When people first came to the Wet Tropics, the tropics weren't wet. The climate was cooler, drier and windier, and there were regular volcanic outbursts. People roamed the savannas in mobile societies, focusing on the riverine environments. Around Thiaki, the rainforest had retreated to the safety of deep river gullies and high mountains: only about 15 per cent of the Atherton Tablelands was rainforest.

As the planet tilted out of its glacial phase, global temperatures increased quickly, melting ice at the poles and warming the glaciers that were squatting on the northern continents. Starting around 15,000 years ago, the Great Barrier Reef shelf began to flood. By 8500[71] years ago, it was completely inundated, which reinitiated coral growth. With the high seas, the monsoon was once again invigorated and, with additional moisture from tropical trade winds, the wetter conditions coaxed the rainforests from their refugial pools.[72] They flowed across the landscape in an expanse larger than pre-European rainforests.

People were quick to inhabit them.[73] And by 5000 years ago, tribes of people lived almost exclusively within the forest. But as the climate once again shifted to extremes, with El Niño droughts and reduced monsoons, the rainforests shifted composition to drier types. This, it seems, galvanised people to develop an extraordinary culture: one based on a staple of toxic nuts. Uniquely found in the Wet Tropics,[74] the nuts were mainly from four endemic species: *Beilschmiedia bancroftii* (yellow walnut), *Endiandra palmerstonii* (black walnut), *Prumnopitys amara* (black pine) and *Castanospermum australe* (black bean).

Naturalist Eric Mjöberg, who visited the area around Thiaki in 1913, described—in vaguely unflattering terms—the plant foods available within tropical Australian rainforests at that time:

[T]his is the time of the year when the rainforest fruits fall down from the trees. The ground looks like an orchard after a . . . storm. All sorts of delicious to look at fruit are found— but bitter to taste, competing in numbers and in wonderful colours of yellow, red, green, and blue. Bright red cherries, an inch across, as bitter as gall; plums, which are not plums at all; false apples, oranges, walnuts and chestnuts, which, now and then, drop down through the foliage. This is harvest time for the blacks [sic] who eagerly search out and find the walnut-like fruits of Cryptocarya [Beilschmiedia] bancroftii. The old women act as beasts of burden. One can see them constantly passing by, carrying their cane baskets, filled with nuts, to be stored and later to be crushed, roasted and thoroughly washed in water, before being eaten.[75]

The process of converting the toxic nuts into nutritious food was intricate and time-consuming. A ground oven was dug, and stones, paperbark, ginger leaves and firewood were collected. The fire was lit on top of the stone-lined oven. The black walnut and black pine nut were cooked on the coals of the fire until their shells cracked; the seeds were removed by prising the shells open or cracking them further with a stone. The black walnut and black pine nut could be eaten directly or, by adding water, they could be made into a paste that was wrapped in leaves and returned to the coals for 10–15 minutes.

The yellow walnut and black bean needed more treatment. They were placed in the ground oven, which had been lined with wet ginger leaves, and then covered with another layer of ginger leaves, followed by a layer of soil. A small fire was built on top using the coals from the original fire, and it needed to be maintained: four hours for the yellow walnut and eight hours for the black bean. The seeds of the yellow walnut were then removed, grated into meal, wrapped in ginger leaves and cooked for another 45 minutes in the ashes. The black bean was sliced into chip-like segments after cooking. The meal and the chips were then leached in a beautifully woven dillybag made from lawyer vine and placed securely between two rocks in a small creek. The yellow walnut was leached overnight, and the black bean submerged in a gentle stream for four days.

'The nuts give you lots of energy you know, you feel like you are full of beans after eating them' was how one Aboriginal Elder described the effects of eating detoxified rainforest walnuts.[76] Amazingly, the heating, pounding and leaching not only rids the nuts of toxins, but also converts resistant starch—most of the starch in the nuts—into digestible starch. Just one of the nuts, yellow walnut, has been shown to have three times the amount of energy as the equivalent amount of potato. It seems that the nuts are a powerhouse. The processing of nuts by the rainforest peoples was a continuous enterprise associated with other food gathering and social activities so that there was always something to eat: so much food, in fact, that there was enough to cache in wet mud or to stockpile in the camps to feed large wet-season ceremonial gatherings.

Wandering around Thiaki, I've occasionally come across patches dominated by groves of food trees, a structure not generally considered a random occurrence. Black bean doesn't occur in Thiaki. Yet. It grows from sea level to around 800 metres, which is just down the road. Here at the edge of their range, many of the paddock trees are black bean. Perhaps they were kept for their showy sprays of yellow and orange flowers or because they are a useful general-purpose timber. Or simply because they had been a dominant tree once managed by people for food.

Early in the wet season of 2022, we planted a new batch of restoration seedlings: hundreds of trees from 16 species, including black bean. All of these species have a broad altitudinal range from near the coast to upwards of 800 metres. Part of the research with James Cook University researchers, our aim is to integrate climate adaptation into our restoration plantings. Earlier plantings came from seeds sourced locally, the idea being that this local provenance would ensure local adaptation. Climate change, however, is now causing maladaptation, and we are testing to see if it matters where the seeds are from. If trees sourced from the warmer lowlands do well higher up, we have boosted our capacity to successfully restore degraded lands and perhaps got the jump on adaptation.

———

Aboriginal engagement with the landscape was complex and specific. Burning was one of many tools used by Aboriginal people. It was applied as an artist applies a brush to a landscape painting, mindfully, carefully and with intimate knowledge of the expected outcome. The idea of broadscale burning on the continental scale is a vast simplification.[77]

The Kuku Yalanji rainforest peoples north of Thiaki maintain elements of traditional Aboriginal rainforest culture. Kuku Yalanji clan estates are about 5000 hectares; and about 20–50 people hunted and gathered over the estate for most of the year. While most of the food plants came from the rainforest and swamp communities, each estate included a small area of open forest, just a few hundred hectares, perhaps as small as seven per cent of their total estate.

Traditionally, staple carbohydrate food comes from three main areas: the cycads of the open forest, the yams found on the margin of the rainforest, and the rainforest nuts. Each staple is managed with a particular application of fire: the yellow walnut is protected from hot fire in any season that would kill large trees; during the winter dry season, the productivity of the yams is enhanced by burning along rainforest margins to prevent any hot fires that would kill the yam crop before it was harvested; in the hot dry times of spring and early summer, a diversity of fire is lit in the open-forest grasses for access to the cycads growing there, to enhance their production and to create a diverse open-forest community that did not eat into the rainforest.

So important are the cycads that stands of them are apportioned among women and bequeathed to daughters or other relatives.

Water is central to the processing of cycads, and women are said to have the power to control the flow of water in special places.[78]

———

What happens to Country when people are taken out of the picture? In the vast Western Desert of Australia, the Martu people were cleared from Country in the late 1960s, severing in an instant their millennia-long patterning of Country with gentle fires that created niches for plants and animals. It wasn't long before large wildfires incinerated the landscape. The rufous hare-wallaby, burrowing bettong, bilby, mulgara and brushtail possum became locally extinct or endangered. Invasive species—rabbits, cats, camels, foxes and donkeys—took over.

When the Martu returned in the 1980s, scientists travelled with them to see if they could map Martu knowledge. Emerging from the ashes was a network of 1149 different links based on 173 edible plants and animals (about 12 per cent of all identified species), including what they ate and what ate them. This didn't include invertebrates, as the level of detail was simply too complex for the researchers to fathom. The scientists found what the Martu had already known— their management strengthened desert ecosystems.[79]

In the Wet Tropics, the population of Bennett's tree-kangaroo is sparse as it's thought to prefer high altitudes. Few were recorded historically. But in about 1960, populations unexpectedly appeared in lower altitudes before once again fleeing up to mountain strongholds.[80] The shift back up to deep-rainforest mountain refuges coincides with climate change. But why did the tree-kangaroo population expand downwards in the 1960s? An understanding of the area's history provides an insight.

In 1874, thousands of European miners swamped Cooktown and rushed single-mindedly to the Palmer River goldfields without any interest in observing wildlife. Rainforest peoples were forced deep into high rainforest refuges, where hunting was usually restricted by Law. Prospector James Venture Mulligan observed that, by 1881, 'the blacks are there in thousands; this is now their stronghold. Having been displaced from the Normanby, the Palmer, the Hodgkinson, Port Douglas etc. they have made the Daintree and Bloomfield their rendezvous, and are determined to hold it as such against all comers.'[81] Squeezed into a small area, Aboriginal hunting restrictions—including of the tree-kangaroo—must have given way as starvation took hold. Eventually, people were forced to 'come in' to European communities for food. Released from hunting pressure, the tree-kangaroo's population may have rebounded back to pre-European levels and it may have been this that people noted in the 1960s. Rather than being confined to mountain tops, Bennett's tree-kangaroo may always have had expanded populations carefully managed by people.

In the Top End of Australia, a pollen record extracted in a core from an ancient lagoon in suburban Darwin[82] tells a story beginning with a cracked and dry landscape at the end of the last glacial period. Monsoons returned about 13,000 years ago with force, only to weaken and then strengthen again. Like a defibrillated heartbeat, the landscape pulsed with the rainfall, each pulse more successful than the last in coaxing wetland tree growth and, finally, a standing pond of water.

The core shows that rainfall increased in quantity and reliability between 9000 years ago and 4000 years ago as the sea rose from its glacial low stand. Rushes and waterlilies flourished in a lagoon fringed by a monsoon forest of *Melaleuca*, *Pandanus*, *Banksia* and *Grevillea*. Higher rainfall allowed non-eucalypt sclerophyll, monsoon rainforest species and moisture-loving mid-layer shrubs and trees such as *Pandanus* to increase, closing the canopy and smothering the fire-starting grass.

But then the climate shifted to one of volatility, swinging unpredictably between monsoons and long drought with the onset of El Niño. However, instead of the expected vegetation of open woody forest dominated by destructive fire, the core shows a stable forest of eucalypts that had shaded out the volatile grasses. Fire was less, not more, intense. And even in the face of increasing drought, the landscape filled with layers of diverse herbs and shrubs and a variety of trees, with groves of monsoon forest closer to the lagoon.[83] It was clear that the patterning of the landscape was the handiwork of people with an intimate and deep ecological knowledge of their Country, working skilfully with fire.

Post-glacial greening allowed the number of people to increase, together with an increase in technical and cultural innovation that gentled fire so that it was not an agent of destruction.[84] Instead, and even in the face of a volatile climate, fire was put to use to enhance biodiversity and buffer the destructive effects of droughts. It seems that these future-makers had almost miraculously decoupled the landscape from climate.[85]

But then, around 750 years ago, the Top End core shows a decline in forest cover when destructive fire was once again released across the landscape. The pattern suggests that the number of Aboriginal people plummeted, perhaps due to disease that might have been brought by Macassans. The effects on the landscape were like those of depopulation after European contact.

———

In the Wet Tropics, lake cores show that dramatic rainforest expansion came with the rejuvenated monsoons and trade winds: close to the coastal ranges, rainforest replaced eucalypt forests from 17,000 years ago, and further west on the Atherton Tablelands, where it was drier, around 8000 years ago.[86] Against this climatic backdrop, the management of fire was particularly fine-drawn,[87] calculated to augment, not diminish, plant and animal cycles with a thoroughness and scale that was breathtaking. The rainforest was traced with pathways and enlivened with open pockets of sclerophyll. The shaping of ecotone boundaries between sclerophyll and rainforest was intricately managed against wildfires, and to protect useful groves of fruit and nut trees.

Throughout the Wet Tropics, the number of sclerophyll pockets are in the thousands. They range in size from 1 hectare to hundreds of hectares,[88] and are often circular in shape and sharply demarcated by rainforest. Woven through the landscape is a network of traditional pathways linking what were significant sites: open pockets in the rainforest—small clearings made by Aboriginal people for camping and ceremony—creek lines, lakes and swamps. Curtain Fig Tree, now a major tourist attraction, was an important meeting place for Ngadjon people when the tree was small. Nearby are the sacred centres of Noopah and Chumbrumba: bora grounds, where people manipulated time and space to open a dialogue between themselves and their ancestors, at the foothills of Mount Quincan, the embodiment of an important ancestor—now being quarried for 'quincan' gravel.

Europeans claimed the chains of pockets as their main pack-supply track through the rainforest, connecting the mining town of Herberton with Cairns. Some pockets became occupation sites for European towns: Allumbah Pocket became Yungaburra, and a cedar-cutter's camp on Prior's Pocket became Atherton.[89] Noopah Pocket, Chumbrumba Pocket, Dingo Pocket, Mooma Pocket and Kirrakopa Pocket were subsumed into grazing land.

Not well known is the fact that more than half of the Wet Tropics bioregion is dominated by non-rainforest communities, which evolved alongside—and are as unique as—the rainforest. Populations of northern bettongs and yellow-bellied gliders, now in retreat, rely on them. Recent meticulous vegetation mapping has shown that individual forest types of the bioregion face unprecedented habitat change due to the cessation of Aboriginal burning. An irruption of rainforest understoreys is occurring across most of the sclerophyll forests and woodlands. While on the face of it this may sound like good news, the fact is that these are a weedy type of forest, poorer in plant and animal diversity than the sclerophyll communities they replace.[90] They are no substitute for the rich, sustainable and stable environments created by the rainforest peoples.

# PART TWO

# People of the Rainforest

# CHAPTER NINE

# Thiaki's theatre of the absurd

Country—animals, trees, rain, rivers, hills, sun and moon—sometimes watches humans and thinks about them.[1] The stars are intimately involved in the affairs of people.[2] In this cosmology, humans are an often-obscure part of the world. Wisdom is the knowledge of relationships between humans and other species, and acting responsibly in that knowledge. Humans know that birds, insects, kangaroos, trees, grass, moss, rain and stars also know that wisdom lies in awareness of life systems and behaving with intention and with responsibility.

Each sacred site bears a name and acknowledges the continuing presence of the Dreamtime. The magnificent beings may be gone or sleeping, but Country is still alive with their Law. Insect, plant, animal, person . . . all have rules: how to get and distribute food; which wood to use for fire; how to make tools and shelter; which scars and marks to wear to carry the ongoing presence of the Dreamtime. Every character has a part to play in the web of life. All forms of life are intimately linked and share the same life essence.

———

In eighteenth-century Europe, the dominant metaphysical view of life was as a great Chain of Being. Decreed by God, with whom the chain started, it descended through angels, humans and then to lowly animals and minerals. Derived ultimately from Plato and Aristotle but manipulated to suit imperial ambitions, the view explained what appeared to be a ladder of nature in which it was very difficult to move up from one's obdurate rung, a difficulty not encountered if you were powerful. In the Middle Ages, for instance, it became convenient for the chain to grow more links to accommodate the inequalities of feudal society. Flowing downward from God were now angels, kings and queens, and then the whole feudal system from archbishops to dukes, bishops to lesser nobles, then commoners, tradesmen, servants, tenant farmers, beggars and thieves.[3] 'Primitive men', or 'savages', were close to the bottom, just above apes, animals, birds, insects, worms, minerals and rock. Among men, Europeans were ranked pre-eminent. Aboriginal peoples just scraped in as human.[4]

The shifting metaphysical foundation of this pernicious idea would ultimately be the undoing of the forests around Thiaki.

Captain James Cook's rather wistful, even envious, summation of people was written against this background of the rigid inequalities justified by the Chain of Being. He found that Aboriginal people were

> a timorous and inoffensive race, no ways inclinable to cruelty, as appear'd from their behaviour to one of our people in Endeavour River . . . in reality they are far more happier than we Europeans; being wholly unacquainted not only with the superfluous but the necessary Conveniences so much sought after in Europe they are happy in not knowing the use of them. They live in a Tranquillity which is not disturb'd by the Inequality of Condition. The Earth and sea of their own

accord furnishes them with all things necessary for life, they covet no Magnificent House, household-staff etc, they live in a warm and fine Climate and enjoy a very Wholesome Air . . . In short they seem'd to set no Value upon any thing we gave them, nor would they ever part with any thing of their own for any one article we could offer them; This in my opinion argues that they think themselves provided with all the necessarys of Life and that they have no superfluities.[5]

This didn't stop him from claiming eastern Australia for England.

―――

The land naturally produces hardly anything fit for Man to eat, and the Natives know nothing of Cultivation.

Captain James Cook[6]

On 21 May 1848, the barque *Tam O'Shanter* in company with HMS *Rattlesnake* skirted northward around the jagged peaks of Hinchinbrook Island, scarfed in cloud, and anchored in the shallow tropical waters of Rockingham Bay in Far North Queensland.[7] Clear, clean and peaceful, the waters were shocked by the slap of a wooden boat as it was lowered. Captain Merionberg and Edmund Kennedy, the leader of the Kennedy Expedition—'undertaken for the exploration of the country lying between Rockingham Bay and Cape York'—clambered into the boat to reconnoitre a safe landing place for the expedition's 28 horses and 100 sheep. Since the explorer Ludwig Leichhardt had found good grazing land in 1844, only about 100 kilometres inland as the crow flies, Kennedy's job was to be a quick and easy reconnoitre of the unexplored coastal land.[8]

Three days of rowing along the shore of the extensive bay had failed to find a safe place to anchor close to shore, so *Tam O'Shanter* was anchored about a half a kilometre out near a rocky outcrop at the northernmost point of Rockingham Bay. Here, the horses might safely swim to shore where grass and fresh water had been found.

*Tam O'Shanter* had anchored in Djiru Country, and some Djiru people had gathered along the beach to watch the weird and wonderful proceedings through the cool drizzle of that day. White men and their boats were not new. From 1770 to 1848, there had been around 13 official explorations along the Great Barrier Reef as well as dozens of undocumented, often dubious, commercial vessels plying the waters—particularly after the 1830s because of the lucrative trade of Waler horses for the British Indian Army. The waters were treacherous and, by 1854, at least 70 ships and 150 lives had been lost in the reef-ripped waters.[9] Bêche-de-mer (sea slug) fisheries for the Asian market were also well established by the 1840s. They frequently kidnapped people to work as slaves.[10]

Even as the *Tam O'Shanter* was earlier anchored at Palm Island, not far south of Rockingham Bay, a shady conflict was taking place on the *Will O' the Wisp*, which was in search of sandalwood.[11] The naturalist on the *Rattlesnake*, John MacGillivray, observed the small cutter beating up the coast 'fitted out by a merchant in Sydney and sent in a somewhat mysterious way (so as to ensure secrecy) to search for sandalwood upon the north-east coast of Australia. If found in sufficient quantity, a party was to be left to cut it, while the vessel returned to Moreton Bay with the news and convey it at once to the China market.'[12] The *Will O' the Wisp* was making a beeline for the *Rattlesnake* in order to seek help for its injured crew and captain.

The crew said that while they were anchored off Palm Island, people came to the ship in their canoes. They were apparently offered gifts, and some came aboard. What they saw or how they were treated must have enraged the people because they raided the ship in the dead of the night. Five or six large canoes with outriggers came alongside at 3.30 a.m. while the crew were asleep, throwing lighted bark into the cabin and down the hatchway. The master and one crew member were knocked unconscious by boomerangs.

The mate, however, got hold of a cutlass and laid about with it. By now, the crew were awake and shot some people who fled. The crew then manned a deadly swivel gun (a small cannon), mounted amidships, and 'the survivors retired in confusion'. The ship fled north, stopping for water on Goold Island, where the crew opened fire on a large body of people from a completely different tribe. The crew said they were hostile, but they were probably going out to meet the vessel as they had all others. Finding the account suspicious, Captain Stanley of the *Rattlesnake* reported it to the Colonial Government. What seems sure is that many people were killed, none of them the crew of the *Will O' the Wisp*.

———

Rockingham Bay people comprised at least five different tribal groups, who were not naturally hostile. In 1843, Joseph Beete Jukes, naturalist of HMS *Fly* and one of the first Europeans known to have set foot on shore in the Wet Tropics, commented that the people of Rockingham Bay were friendly and very numerous. He often found himself surrounded by different groups of 40 or 50 people[13] as he explored along native tracks and through belts of mangroves surrounded by open grassy country that stretched out as far as he could see. Adjacent to the creek he found swamps full of fish.

For a people who adored theatre, the *Tam O'Shanter* show must have been thrilling. Carrying green branches, like admission tickets, to indicate peaceful intent, the Djiru advanced closer. The white men, in turn, presented to 'fine strong young men' two circular tin plates with Kennedy's initials stamped on them and chains to hang around their necks. Of more interest perhaps were the gifts of fishhooks.

A line was drawn in the sand by one of the intruders, beyond which the Djiru were asked not to step. In so many ways, this line was a profound metaphor for the rainforest peoples. But for today it was just another strange request from a very strange folk indeed.

Politely, the Djiru sat at the line to debate among themselves and look on in amused amazement at the odd performance being acted out before them on the beach.

A sling hoisted a horse from inside of the vessel into the air and then down into the water alongside a small boat to which the horse's head was secured so the animal could be towed to shore, the whites of its eyes bulging in panic. As the day progressed, 26 more horses were aerially launched and lowered into the sea, with one poor creature drowning in thrashing terror. Touching down on the beach were also 100 sheep (together with a sheep pen), three kangaroo dogs and a sheep dog. Inanimate objects also began to stack up: 1 ton of flour, 90 pounds of tea, 600 pounds of sugar, 24 pack saddles, one heavy square cart, two spring carts with harnesses for nine horses, four tents, a canvas sheepfold, 22 pounds of gunpowder, 130 pounds of shot, a quarter cask of ammunition, 28 tether ropes (each 21 yards long), 40 hobble chains and straps, together with boxes, paper for preserving specimens, firearms, cloaks, blankets and tomahawks.

Canvas tents were pitched a short distance from the beach and a few short steps from a small freshwater creek. One large tent, for Mr Kennedy, had a foldable table, stool, books and instruments. In the evening, every expeditioner was issued a double-barrelled gun, a brace of pistols, a cartridge box and rounds of ammunition. Expeditioners then gathered on a beach thickly strewn with shells—the remnants of thousands of meals—to eat beef and biscuits.

The Djiru returned to their camps, perhaps planning the hilarious re-enactment they would present to their folk after a meal of fresh fish and shellfish. They may also have eaten a wallaby, some fresh greens, fruits, nuts and roasted yams.

Chafing at the bit, William Carron,[14] botanist of the expedition, broke the barriers of the camp early the next morning for a spot of botanising. He headed north towards the nearby rocky outcrop to look at the glorious orchids he'd spied there. Walking through a small belt of scrub, he marvelled at the diversity of vegetation, then

Morning mist over the forest on our property, Thiaki, in the Atherton Tablelands. Bellenden Ker Range is in the distance.

Life on Thiaki. The author with granddaughter Milla, Charlie the cattle dog, and cattle. Restoration plantings are in the background.

Preparation for restoration on Thiaki in 2011. Rows are ready for planting.

Maturing rainforest plantings in 2022. This photograph is taken from the same site as the image above.

Newly planted seedlings with our home in the background.

A golden bowerbird, one of the endemic wet tropical rainforest birds we see on Thiaki. It is vulnerable to climate change, and with 3 degrees Celsius warming the golden bowerbird is predicted to go extinct. (Photo Martin Willis)

Walking down into platypus creek, with its remnant original rainforest. Planting is in progress on the far hillside.

Planters at work on the steep slopes of Thiaki. We are experimenting with planting different mixes and numbers of rainforest tree species with different planting densities to see how to build a biodiverse rainforest from a paddock most cost-effectively.

A magnificent red cedar, 400–500 years old, covered in epiphytes, at Mossman Gorge. Cedar-cutters stripped the north Queensland rainforests of mature cedars like this in the late nineteenth century. (Photo Elizabeth Weiss)

The glorious wompoo fruit-dove has a strange, deep-throated *whuck woo* call I mistook for 'fuck you'. (Photo Martin Willis)

*Austrobaileya scandens*, one of the most ancient of flowering plants, grows near our back deck.

*Dawsonia*, a large moss with a stem of up to 20 centimetres, carpets sun-spilt gaps on the forest floor at Thiaki.

Silvereyes are among the fastest evolving of all terrestrial vertebrates. At certain times of the year, they fly in flocks around the property, constantly calling *pee pee*. (Photo Martin Willis)

The green ringtail possum we see on Thiaki is found only in the Wet Tropics. Four or five of its close relatives migrated north to New Guinea as the rainforests of Australia shrank and before New Guinea was separated from Australia by the Arafura Sea.

The Norwegian explorer Carl Lumholtz spent ten months in north Queensland in 1882 and described the tree-kangaroo named after him as the most beautiful mammal he'd ever seen. We see this tree-climbing descendant of the rock-wallaby regularly around Thiaki. (Photo Noel Preece)

Thiaki is a known hotspot for the Herbert River ringtail possum (pictured) and the lemuroid ringtail possum (below). (Photo Martin Willis)

The lemuroid ringtail possum. (Photo Martin Willis)

A family group on the Atherton Tablelands in the 1890s with characteristic rainforest items: a bark basket for carrying water (left) and a bicornual basket made from lawyer cane (right), as well as shields and spears. (Courtesy Eacham Historical Society)

Historical photograph of an Aboriginal group from Bellenden Ker, with shields, spears and boomerangs. (Courtesy Eacham Historical Society)

Traditional Owner of Thiaki, Ngadjon man Ernie Raymont (left), and his grandmother Molly Raymond (right), who was the last Ngadjon person to be born in the rainforest. (Courtesy Eacham Historical Society)

Native police with 'camp followers' and white supervisor in the Atherton Tablelands. Queensland's native police were known as the most efficient and well-trained killing machines in the world. (Courtesy Eacham Historical Society)

A young man climbing a tree with the use of rope made from the *Calamus* vine. People hunted at least 20 different types of animals in the canopy. (Courtesy Eacham Historical Society)

Cedar-getters worked their way from Ulladulla, south of Sydney, in the 1820s, to northern Queensland, erasing the red cedar from the landscape by the end of the century. Other valuable species, such as this spectacular bloodwood, disappeared along with them. (Courtesy Eacham Historical Society)

A bullock team hauling a log through a creek. Logging was often lawless, a wholesale system of plunder. Only the best part of the tree was cut and squared, leaving most tree limbs in a trail of destruction through the forest. (Courtesy Eacham Historical Society)

A huge log with the size chalked on its stump. Red-cedar trees could grow to 60 metres high with girths of 11 metres. (Courtesy Eacham Historical Society)

Tree felling on the Atherton Tablelands in the 1890s. The Atherton Tablelands were some of the last accessible rainforests of Australia to fall in the pursuit of the fantasy of a yeoman's paradise. (Courtesy Eacham Historical Society)

A bullock team hauling logs against a background of cleared land with stumps. By the 1980s, nearly 80,000 hectares of forest had been clawed away on the Atherton Tablelands, leaving just 100 rainforest fragments in poor health and with dwindling native animal populations. (Courtesy Eacham Historical Society)

Newly cleared land around Malanda, circa 1908. Steep hillsides were stripped of rainforest to create grazing land for dairy cattle, but the soil rapidly lost fertility once the trees were gone. (Courtesy Eacham Historical Society)

Thiaki's restored slope in the foreground, showing how the rainforest catches the mist, a phenomenon known as cloud stripping. In the background is the Herberton Range, part of the Wet Tropics World Heritage Area.

edged around a salty mangrove swamp and entered a freshwater swamp ringed with *Melaleuca* (tea-tree). Nearly all of the plants Carron observed were important to the Djiru. The leaves of *Melaleuca*, for instance, were used as an antiseptic. He noted the loops, metres-high, of strongly growing *Calamus*, a climbing palm used for innumerable purposes from making huts to baskets. Between the beach and swamp were stately flooded gums; large, shady, spreading *Cerbera* species (milkwoods); *Dillenia alata* (red beech); and the gorgeous *Calophyllum inophyllum* (beauty leaf), widely cultivated throughout the tropics,[15] spreading out its inviting shade towards the sea.

*Flagellaria*, a lawyer vine that has no prickles, was used to make baskets and nets, as well as for food and medicine: its growing tips were eaten, and the stinging sap from the young plant was used on sores.[16] Bundles of dry vines were also set on fire and used as a torch during night hunting. The common *Smilax* vine has leaves and ripe black fruits that could be used as a medicine for coughs and chest complaints.[17] *Cissus*, a native grape, was eaten raw. The *Cissus* vine itself is strong enough to make rope for climbing trees. *Eugenia reinwardtiana* (beach cherry) has red fruits which are sweet and edible.

Carron frequently found *Castanospermum australe* (black bean). People ate the nuts (processed so they were no longer toxic); the bark fibre was pounded and used for fish and animal traps, nets and baskets; and the wood was used for spear-throwers. The tree itself spoke to the people as a seasonal cue for hunting jungle fowl. So important was black bean that it was almost certainly carried along the eastern coastline as far south as northern New South Wales, where it is found today growing in the Clarence and Richmond river catchments. An important songline tells of an ancestor who carried the black bean along the ridges dividing New South Wales and Queensland. The genetics of the black bean populations suggest that they are closely related to one recent ancestor.[18]

The handsome tree fern *Dicksonia* was important, for the starchy pulp of the top of the trunk was roasted. The *Dendrobium* orchids that caught Carron's eye had their stems beaten to break the fibre before being cooked and eaten. In the freshwater swamps behind the dunes were dozens of freshwater fish species, turtles, frogs, snakes and other reptiles. *Exocarpos*, in the sandalwood family, was used for making spear-throwers and bull-roarers, the bark was a contraceptive and the leaves were burned to repel insects or, in a solution, used to treat sores. The seeds of the succulent *Portulaca* were used to make bread. The grass tree *Xanthorrhoea* was used for glue, and *Lomandra* was used to make baskets.

The land was beautiful and wildly luxuriant. And that was no accident. It was a skilfully managed and bountiful garden worked by a multitude of guiding hands; the resources were conveniently available and predictable, thanks to an intergenerational eye for detail and a deep ecological knowledge of fire, wind, fuel loads and future resource needs.[19] The open ground between the beach and the swamp, varying in width between 0.8 and 6.4 kilometres, was covered in long grass in which lived a great many wallabies. An intricate mosaic of rainforest, open eucalypt forest, swamp and coast produced a plentiful diversity of life.

In the dunes behind the beach were fruit trees interlaced with grassy swathes for game; the swamp edges were managed for grass, forest or *Melaleuca* (tea-tree). Bananas, yams, fruits and nuts were encouraged to grow where it best suited them. Honey could be licked fresh from grevillea flowers or concentrated by different native bees—the small yellow forest bee having the sweetest. There were many types of juicy grubs and crunchy beetles, depending on the tree or bush. Snails could be roasted in their juice, and sieved termite nest was a specialty cooked in aromatic leaves. Fern fronds and mushrooms were eaten fresh. Other vegetables were better roasted, such as vines and certain leaves, delicious waterlily bulbs and orchid bulbs, carrot-like roots, different types of yams, and other roots such as potatoes, a local wild taro and the beans from the mangroves. Dozens of different plums and berries could be plucked and eaten while gathering. These include Davidson's plum,

Burdekin plum, bush apples, lilly pillies, the sour native tamarind and native cherries. The bush guavas were cautiously eaten, since they burn the mouth. Porridge could be made from a certain type of yam, and cabbage from palm tree heart and gingers were used to flavour food. Breads and cakes were made from cycad and zamia fruits tossed down with the sour juice of green ants.

If you'd told him, Carron would scarcely believe that he was looking upon a managed estate. He would almost certainly have dismissed the idea that the plants, like everything else in the world, were kin and conscious and needed respectful management.

The Kennedy Expedition, comprising 13 men, had a weekly ration of flour, sugar and tea. And a sheep every other day. Their horses and sheep grazed on grass managed by people for several species of wallabies. Bettongs, rat-kangaroos, possums, gliders and bandicoots were in abundance. Sea turtles were occasionally caught and cooked in their shells by the local people. Several types of goannas, dragons and pythons were relished. Geese, ducks, turkeys and a variety of waterbirds were deep buried and slow roasted in their juices. Seafood was almost certainly on the menu. In the sea were bream, grunter, mangrove jack, barramundi, catfish, jewfish, cod, stingray, eel and many types of shark. Prawn, crayfish, oyster and mussel shells were piled high on the beach where people sat, contentedly snacking.

In a land of almost absurd profligacy, Carron dug a piece of ground and sowed seeds of cabbage, turnip, leek, pumpkin, melon and pomegranate as well as peach stones and apple pips.

People knew Country in minute detail: every river loop, certain trees, pockets of grass, where certain animals lived, fallen trees, sandbanks, slopes. Many places were totem sites; some were acknowledged by all people within a wide surrounding area as being of religious significance, while others were of special signifi-cance to particular families or individuals. A cluster of these places comprised a clan's Country, the totem sites in effect stating a claim of corporate title. But one that included not only the possession of the place but also 'the mystical relation with its makers . . . rights over human possessors, natural products and forms of immaterial property' such as the songs, myths, dances, ceremonies and magic potency linked with it.[20]

Camps and more permanent villages were placed with easy access to plant and animal food. Tracks led from one abundant swamp to another. Grass also harboured edible bulbs that at certain times of year erupted to clothe the earth in flowers. It was a pattern repeated, almost like a template,[21] up and down the captivating coast.

———

Kennedy had reconnoitred to the north-west and found swamps and thick scrub. It was singularly the most difficult part of the northern coast to land the ridiculous wheeled carts and farmyard beasts. The 13 men and their clutter of impedimenta were ensnared by swamps and dense scrub fed by rivers tumbling down the mountain rampart directly to the west. Ironically for Kennedy, today the area is celebrated as the nationally significant Edmund Kennedy Wetlands, with its lowland rainforest, paperbark and sedge swamps, and extensive mangrove forests that include most of the mangrove species found in Australia.

On 5 June 1848, after days of preparation, the expedition harnessed horses to carts, saddled and loaded the packhorses, and set off. Kennedy and an Aboriginal guide, Jackey, were at the front, followed by the carts and horses, then a flock of sheep and finally Carron with a barometer.

They travelled perhaps 2 kilometres before they reached the Hull River, which was about 100 metres wide at the mouth. It marked the boundary of the Djiru and Gulngay Nations. The *Rattlesnake* sent a boat to help the expedition across, while a large crowd of bemused Gulngay and Djiru examined the luggage. The next day, 6 June, the *Rattlesnake* departed, leaving the bumbling expedition to lurch and stumble, dragging and hauling its gear preposterously through the sand. By dusk they were halted by another river, the Tully, a further 4 kilometres down the beach. It was about 300 metres wide at its mouth. Mangroves grew on the north side of the river, with grass and scattered *Casuarina* and *Acacia* on the other. Exhausted, the expeditioners slept on the sand.

People had by now congregated in increasing numbers to watch the comical performance. And they were not disappointed. The Europeans constructed a peculiar punt by taking the wheels and axles off one of the carts, which was then wrapped in a tarpaulin. Floats of water bags filled with air were fixed to the sides of the punt, and empty kegs were attached to the shaft. In this manner, most of the things were moved across the river. The next day saw the crossing complete, with the horses feeding on palatable grass meant for wallabies. Kennedy and Jackey, who had spent the day trying to find a better route, returned with the news that they must continue along the beach. Some 6 kilometres further along, they reached another river (the Murray). By now they had passed into lands of the Girramay Nation. It was obvious that the land was closely settled, fertile and productive.

Nearby was a group of 20 oval huts of varying dimensions, the largest being 6 metres long, 2 metres wide and 4 metres high. They were neatly and strongly built, with small saplings dug in and arched over, tied together at the top with a strong rope made from *Calamus*, and roofed with slabs of bark fastened with rope. The floors were strewn with long dried grass for bedding. The village was perfectly placed on grass at the head of tidal waters and close to the freshwater swamp. Pleasant and convenient, it had access to an abundance of resources.

Carron was impressed by a strong and beautifully designed shield plus four hardwood swords that he found much too heavy to wield with one hand. There were a number of fishing lines made from grass, with hooks of various sizes made beautifully from shells. In the centre of the camp were four large ovens about a metre wide and slightly less deep, filled with stones to within 15 centimetres of the surface. These stones would be heated by a fire built on top, and the seafood, wallaby and many of the vegetables would be baked on the stones. Several large shells held water, and there were bottle gourds from a vine growing plentifully along the beach that allowed 1–2 litres of water to be carried.

The next day, some local fishermen in bark canoes aided the Europeans in crossing the river. Some of them clearly had never seen horses or sheep, and their dogs rushed the sheep and drove them in all directions.

The expedition moved several kilometres inland and camped near a freshwater swamp while Kennedy tried to find a way around the swamps. They found a small white opossum with very small black spots that might have been the now possibly extinct white lemuroid possum.

The party travelled around 10 kilometres over the next few days, frequently becoming bogged. Closely observed by 80–100 Girramay, on 22 June, they returned to the beach where travel was easier. On 25 June, they came to another small waterway where they saw many cranes and a pheasant coucal. Some Girramay travelled with them. No doubt keen to move the slow, grotesque and destructive party on, they pointed out the best places to cross what is today the Meunga Creek just north of Cardwell. Here, on 27 June, the group finally moved inland and onto higher passable land. They travelled westward through rainforest diversified into patches of open grassed land with stands of *Corymbia tessellaris* (Moreton Bay ash), a type of eucalypt. Carron also found *Musa banksia* (native banana) and a fragrant tree that was likely *Syzygium forte* (white apple), both with edible fruits.

On 28 June, two of the expeditioners were taken very ill. The group halted in an open patch of forest with plenty of grass. It was no doubt named, revered and carefully managed by the Girramay for grazing marsupials. Kennedy ranged forward, looking for a path on which to pull the carts. Returning on 1 July and finding the men still sick, he was forced to stay put. By now the Traditional Owners—with obligations to Country—must have been anxious to move the impolite and voracious invaders on, and some Girramay tried by making threatening gestures with their spears. They were fired on. Four fell.

On 6 July, the expedition members continued their journey, ludicrously lowering their carts with ropes and pulleys down the banks of steep rainforest creeks that ran into the Tully River. Over the next week, they travelled just a few kilometres through rainforest intersected by small rushing creeks; here and there were patches of open grass. Carron found a large quantity of black bean nuts being soaked in the water. He was told by the local people, who were never far away, that they soaked them for five days before cutting them into thin slices to dry in the sun. They then pounded the slices and baked them on a flat stone.

On 14 July, the axle of one of the carts broke. The horses were loaded with spare pack saddles. Another creek saw the carts and some horses bogged in mud. The load was then carried over the men's shoulders while they crossed creeks, plunging in mud up to the knees. After the carts were lifted out by brute force, the exhausted men stopped. On the floodplain of the Tully, they struggled through large swamps of *Melaleuca*. They were on the boundary of the Girramay and Jirrbal Nations, and groups of these people looked on with curiosity—even today, they have stories of the expedition teetering about in tea-tree swamps.

Finally, on 18 July, Kennedy decided to leave the carts and some equipment behind: a specimen box, a cross-cut saw and pickaxes. The Girramay and Jirrbal must have been nonplussed by the scene of 26 horses, followed by sheep and one of the men, who—elated at having ditched the absurd carts—exuberantly sounded a horn. The expedition travelled a further 10 kilometres through open scrub and then started cutting through dense riverine rainforest. After tortuous progress crisscrossing steep, fast-flowing, clear-watered rainforest creeks, a horse was shot. Keeping tabs on the progress of the oafish expedition, some Aboriginal people visited the camp, but they didn't stay long.

On 25 July, at the Tully River, the expeditioners were little more than 50 kilometres away from their original landing site, a distance easily travelled by people in a day or two along well-used tracks. The intruders crossed the river, with several of the horses falling. In sheer exhaustion, the men slept next to the horses, which were tied up in dense forest. The next day, they cut their way to the top of a hill that had a carefully managed open grassland. A horse fell into the river and died. The hilltops in the area were all carefully manicured, clear of forest, dense with grass—deliberately and usefully arranged.[22]

At the foot of the Cardwell Range, deep within the sheer Tully Gorge, on 29 July, Kennedy and Jackey went to find a way through; Jackey guided them along a path that avoided the cliffs and presumably followed local roads that had carried fleet-footed traffic for thousands of years. Carron spent some time botanising, becoming enamoured with the orchids and hibiscus (which had important uses as a medicinal plant). He dug a piece of ground and, once again, in a country of plenty, sowed seeds of cabbage, turnip, rockmelon, watermelon, parsley, leek, pomegranate, cotton and apple.

On 3 August, they made their ascent up a spur to a well-kept grassy flat. At this point, Kennedy realised that the creeks running down from the range formed the rivers and swamps behind Rockingham Bay. Four of the party began clearing forest up the range, while others were looking after the sheep, horses and stores. Carron—alone for a time while minding some gear—was approached by Jirrbal people. He nervously shot a magpie dead to show the effect of the rifle. The people fled.

On 7 and 8 August, in constant rain, the expedition slowly blundered up the range. At night they slept with the horses and found themselves covered with small leeches. The next day, they smashed through a typical notophyll vine forest like that of Thiaki.

——

One day, while discussing the route with Ernie, he said that they must have come across an important local trading path linking the savanna people in the west with the coastal people in the east, and the Ngadjon rainforest folk living in between. This would explain why such a clumsy expedition only took one day to traverse the Atherton Tablelands rainforest to reach the savanna, out of the gaze of Thiaki.

They might have been out of sight, but they were not out of mind.

———

The expedition is remembered because of Kennedy's death by spearing in Cape York. The failure of the expedition is also blamed on the hostility of the people, but they were responsible for one death only—that of Kennedy. While many of the expeditioners died from starvation, the hostility of the Cape York peoples was real enough. And—from Thiaki's point of view—that was welcome because it would at least slow the billowing blast wave of European civilisation heading their way.

Unlike the rainforest peoples, the Cape York peoples had a long history of often violent dealings with invaders from across the seas. Pottery-bearing Melanesians were trading with Torres Strait Islanders and Cape York Nations perhaps as long ago as 3000 years.[23] The Portuguese and Spanish were probably sailing through Torres Strait before 1530, the knowledge being kept secret from other colonial powers.[24] Documented encounters with Europeans were rarely friendly, so it is no surprise that the Cape York peoples were already on guard when, in 1606, Dutchman Willem Jansz of the *Duyfken* came looking for commodities and slaves.

Known for ferocity, ruthlessness and bearing a God-given right to rule the world, the Dutch sent sailors ashore to initiate trade and—where possible—to kidnap, near the current site of Weipa. However, they had not counted on the fearlessness of the Wik people.

In common with Aboriginal society generally, they saw controlled armed conflict as an important, perhaps pivotal, social activity.[25] Known by other Cape York peoples to be strong, politically assertive and populous,[26] the Wik were not to be meddled with. They were moulded by a volatile wet and dry climate that concentrated resources in certain sacred areas. And these were demarcated and fiercely defended from strangers and countrymen alike, who did not have permission to use them.

Along the coast, and inland, lay birth and cremation sites, wet- and dry-season residential sites, trading sites and hunting sites. There were focal sites for food, such as geese nests that dimpled the shell-grit ridges behind the shoreline; lily-rich wetlands; rivers of fish; grasslands of grazing macropods; heavy woodlands; and rich jungles flanking the rivers. The land was infused with totemic sites for shark, salmon, barramundi, rainforest fruit, waterlily, taipan, magpie goose, rat, brolga, dog, honey, storm, rain and wind. It would be impossible for strangers to put a foot down on these supercharged estates without blundering into sacred sites.

Jansz's expeditioners wielded metal implements to dig for water in the sentient ground and guns to shoot totemic wildlife. The Wik may have made initial polite overtures just in case the strangers had come from the sky.[27] And they may even have helped them to find water and food. But European arrogance and attitudes of superiority quickly became a nuisance. Jansz reported that nine Dutchmen were killed 'by the heathens, which are man-eaters'.[28] The number of people killed is not known but is hinted at in the journal of Jan Carstensz, captain of the *Pera*, who undertook a follow-up expedition in 1623. He noted that the hostile Wik had clearly learned the deadly cost of muskets from the *Duyfken*.

Like Jansz, Carstensz was instructed to find valuable commodities and to forcefully steal adults and children. It is worth keeping in mind that both men were servants of the Dutch East India Company, which had massacred most of the 15,000-strong population of the Banda Islands in 1621 for nutmeg and mace. With few people left to do hard labour, the Dutch were now badly in need of slaves. So, on the immaculate Cape York beaches, fringed by a turquoise sea, they held up gleaming metal and sparkling beads as a lure. Those who fell for the ruse were brutally lassoed around the neck and beaten insensible so they could be forced, like hooked marlin, into the boats.

On Cape York, the *Pera* made more than a dozen landings along the Wik coastline, but often faced up to 200 people lining the shores, ready to attack. The same reception probably greeted the next Dutchman, Abel Tasman, in 1644 when he journeyed down the west coast of Cape York, having met with 'naked, beach-roaming wretches destitute of rice and not possessed of any fruits worth mentioning, excessively poor and in many places of very malignant nature'.[29]

The Dutch got the message—for a time. But a century later, in 1756, two boats—the *Buijs* and the *Rijder*—once again set off from Batavia. They became separated in a storm. A boatload of sailors from the *Buijs* went looking for water and firewood off a sandbank in northern Cape York and were never seen again. The *Buijs* turned tail and headed back to Batavia. The crew of the *Rijder*, meanwhile, had sailed further south along Cape York and, in a quest to appease the local people, provided sugared arrack (an alcoholic drink) to a group of Aboriginal men. Amicable relations lasted for several sodden days until the Dutchmen tried to kidnap two young warriors. The Wik warriors fought back with such ferocity that, even though the Dutch felt they had the God-given right 'to exploit all lesser breeds', these men were nevertheless respected for their aggression. The Wik were finally deemed too dangerous for any meaningful relationship to be established.[30]

———

Coming from the west, the Dutch had apparently not found a way around the island-strewn tip of Cape York, and they assumed that

New Guinea was attached to the Australian mainland. In 1770, Captain James Cook proved that wrong once and for all when he sailed up the east coast. He famously holed his boat on the Great Barrier Reef on Monday, 11 June, and limped to the shores of the land-owning Guugu Yimithirr at Endeavour River, now the site of Cooktown.

On 6 June 1770, Cook had passed and named Cleveland Bay (the site of Townsville) and Magnetic Island. On Friday, 8 June, he named Palm Island for what he thought were coconut palms but were, in fact, cabbage palms. He also named Rockingham Bay after the Marquess of Rockingham, who was the British prime minister from 1765 to 1766. On Saturday, he named the Frankland Islands and Fitzroy Island. On Sunday, the *Endeavour* hauled around Cape Grafton and, skirting one of the most majestic rainforest-cloaked mountain backdrops in the world, Cook met nothing to comment on. He continued under sail past the site of Cairns, reluctant to waste the light of a full moon, and passed a coastline glimmering here and there with camp fires. Standing on the deck of the *Endeavour*, naturalist Sir Joseph Banks did not realise that he was passing astonishing tropical rainforests running to the sea, and said he found the eastern seaboard rather desolate and barren.[31] They passed Low Isles and named Trinity Bay between Cape Grafton and Cape Tribulation, near to where the *Endeavour* was holed.[32]

While it only took 11 days to fix the boat, strong annual trade winds were in full force, and the *Endeavour* had to stay put for nearly seven weeks. Unlike the Wik, the Guugu Yimithirr were not familiar with Europeans and waited nearly a month before approaching the strange and frightening party: perhaps they were the returned spirits of the ancestors who had come from the west in the form of the masked plover and wallaroo, in the process turning white?[33] Through careful observation, however, and noting many disrespectful actions, it was determined these apparitions were not their ancestors. The rainforest people finally approached. By all accounts, good relations were established. The Europeans were interested in the people and wrote down 130 words of the Guugu Yimithirr language, including the word 'kangaroo' for an odd creature that bounced rather than ran around.

Conflict ensued, however, after the local men went aboard the *Endeavour* and found 12 turtles that in all likelihood came from a sacred breeding place on the reef. This was an astonishingly greedy number of turtles. It was unthinkable not to want to share them. The next day, ten unarmed men came and formally asked for two of the turtles. They were refused. In anger, the grass around the European camp site was set on fire. Cook, in turn, shot a musket and hit one of the men in the shoulder. This spilled blood on a sacred place: the *Endeavour* had beached on special neutral ground for certain communal activities such as birthing, and bad blood was never to be spilled there.

The fire lighters ran into the bush with Cook in pursuit, and he picked up some dropped spears as he went. Suddenly, an old man appeared from behind some rocks carrying a spear with the tip broken to indicate peace. Cook, understanding the gesture, handed the spears to him in an act of reconciliation. Ten days later—the two groups barely communicated in that period of time—Cook set sail. Behind him, the hills around the Endeavour River were set alight, presumably to cleanse the area of corruptness.[34]

Ships kept coming, though. After the mutiny of the *Bounty* in 1789, Captain William Bligh stayed clear of the coastline, as did the survivors of the *Pandora* shipwreck, in 1791, which included crew members and captured mutineers. Matthew Flinders used the outer route to sail around the Great Barrier Reef in 1802, in the *Investigator*, on his circumnavigation of Australia nearly a century after the Dutch explorations, without incident.

In 1815, the brig *Kangaroo*—an armed transport ship with a regiment of soldiers heading for the British Crown colony of Ceylon—was forced to sail close to the coast due to bad weather. At Goold Island, in Rockingham Bay, friendly Aboriginal people supplied the crew with fruits. Unfortunately for the locals,

Goold Island soon became a well-established anchorage where wood and water were obtained.[35] Other vessels—whalers or sandal-wood traders—plied the waters for illicit reasons, leaving no formal record.

In 1819, Phillip Parker King returned to Australia after the Napoleonic Wars and was assigned to survey the waters along north-eastern Australia that remained mysterious and dangerous. From May to July, he captained the *Mermaid* and noted fires along much of the grassy and wooded eastern coastline. Wherever *Mermaid* anchored, there were clusters of Aboriginal huts and walking tracks through grass and woodlands; further north, along the Wet Tropics coastline, were canoes with outriggers. King demon-strated the use of scissors (for a haircut) to the intrigued people of Palm Island.

He named Hinchinbrook Range and, while not checking, sus-pected that the peaks actually formed an island. At Goold Island, several canoes came alongside, and the people ventured on board to accept snacks of ship's biscuits. They were given gifts of fishhooks and lines, which they reciprocated with neatly made baskets and turtle eggs. When they left the ship, they pointed to their camp and invited the crew to visit. Taking up the offer, some officers went ashore; the men of the tribe—after having carefully sent the women and children away—entertained the crew quietly and peace-ably. Along Mission Beach, King found the country diversified with wooded hills and sandy bays. The sea was teeming with life, and the vessel was steadily followed by a humpback whale: 'a fish which appears to be numerous on all parts of this coast within the reefs'.[36] It was on this voyage that King named Mount Bellenden Ker after a private orchid grower.

In June, the *Mermaid* anchored in Endeavour River for a fort-night. After a week of amicable relations, a fracas ensued: some men, watching the crew of the *Mermaid* washing some clothes, asked to be given them and insisted that one of the sailors should denude himself.[37] A spear was thrown, and shots were fired without injury.

Tempers settled and, several days later, King sent a boat up the Endeavour River to the limit of its navigation. Allan Cunningham, the ship's botanist, noted the dense mangroves and, away from the banks, abundant stands of *Hibiscus tiliaceus*. Unbeknown to him, these were used by skilled craftsmen to make highly valued, strong, elastic ropes of over 60 metres, but weighing only a couple of kilograms.[38] The country further inland was grassy, with eucalypts and cycads. Cunningham—perhaps acting on Cook's ill-observed observation that the land produced nothing to eat—sowed peach and apricot stones.

By now the Guugu Yimithirr must surely have been impatient for the resource-using intruders to go, particularly given that this was not the only vessel plying the waters along the coastline. This fact became obvious when the wreck of the *Frederick*, a merchant ship, was noted further north by the *Mermaid*, on the Flinders Group of islands near Cape Melville.

When the *Mermaid* once again visited the Endeavour River in 1820, taking in water and wood, and exploring the land in a radius of almost 10 kilometres,[39] it became clear that they had outstayed their welcome. The Guugu Yimithirr began furtive attempts at ambush, such as approaching peaceably enough, apparently without spears but surreptitiously dragging them along the ground with their toes.

But the ships kept coming. In 1821, the survey ship *Bathurst*, which had replaced the *Mermaid*[40] and was accompanied by the merchant ship *Dick*, sailed along the east coast of Cape York. They observed other merchant ships, such as the *San Antonio* near the Frankland Islands, offshore of the Mulgrave River. The *Bathurst* landed north of the Endeavour River at Bathurst Bay, on 22 June and was threatened by a group of, perhaps understandably, jaded Aboriginal people on one of the Flinders Group of islands.

The *Bathurst* documented the magnificent—now internationally significant—art of the Flinders Group of islands, which 'represented figures of sharks, porpoises, turtle, lizards (of which there were several seen among the rocks), trepang, starfish, canoes, water-gourds and quadrupeds probably intended for kangaroos and dogs. The figures besides being outlined in dots were decorated all over with the white pigment.'[41] They left several days later for Torres Strait after taking spars, iron bolts and teak planks from the wreck of the *Frederick*.

In 1841, four years after a young Queen Victoria came to the British throne, the *Beagle* under the command of Captain John Lort Stokes sailed around Hinchinbrook Island, observing the number of fires along the coast and concluding that in this part of the conti-nent the inhabitants were numerous and the soils fertile.[42] They stopped at Restoration Island and noted that it was temporarily occupied by a large party of Torres Strait Islanders who had come in canoes. Trade between Torres Strait Islanders and Australian Aboriginal peoples had long been conducted via a maritime canoe network along the entire southern coast of Papua New Guinea, across the Torres Strait and down into the east coast of north Queensland at least as far as Cairns. One of the most popular items was a bamboo smoking pipe.[43] Shell pendants also made their way down the coast, with nautilus-shell bead bands and spear-throwers with baler-shell handles making their way north.

In 1843, the *Fly* and *Bramble* concentrated their attention on the Rockingham Bay area, which was becoming a focus for land explorations, as the first wave of pastoralists were spilling into Queensland. They anchored for several weeks at Goold Island, a known watering place for vessels. Unsurprisingly, the unwanted close attention that the Europeans paid to Country—the penetration of the channels, creeks, rivers and secluded sacred areas—resulted in some Aboriginal hostility. The Rockingham Bay peoples, who for 70 years since Cook's journey had been remarkably tolerant, were to fight back against the ongoing invasion of their lands and gain an unrivalled reputation for ferocity, massacres and cannibalism.[44]

In 1848–9, after having seen off the Kennedy Expedition, HMS *Rattlesnake* continued to closely survey the coastline of the Wet Tropics. On board was John MacGillivray, the ship's naturalist, who discovered a splendid new bird-of-paradise, the riflebird, which was subsequently named *Ptiloris victoriae* after Queen Victoria. On the mainland, MacGillivray made an excursion in a boat up a small creek, probably today's Maria Creek. Several Aboriginal people followed him along the bank. They appeared friendly with no weapons in sight and ventured close enough to accept ship's biscuits. Several neat and roomy dome-shaped huts were built near the mouth of a creek, with fishing nets lying around. Now clearly familiar with firearms, the locals fled in panic when they heard the report of MacGillivray's gun while he was shooting (wildlife) on shore.

The *Rattlesnake* continued its survey of the entire inner route of the Great Barrier Reef between Dunk Island and Booby Island of the Torres Strait. Here, it was arranged that they would meet the Kennedy Expedition at the beginning of October 1848. But the month passed without sign of Kennedy. On the *Rattlesnake*, 31-year-old Oswald Walters Brierly, the ship's artist, took advantage of the time to befriend the people. He took frequent walks in the area with the Kaurareg people, who were keen to show off their Country. Brierly—whose activities were considered 'Niggerizing' by the crew of the ship[45]—noted the bands of scrub interspersed with grass aligned along a swamp. The Country was crafted by fire, with the proximity of grass, swamp and bands of trees along the edges of the flats forming kangaroo traps where the people would lie in wait in the scrub.[46] Thick rich jungle skirted bands of open forests of *Melaleuca* and *Pandanus*. These shady forests were mostly open underneath, with many trees full of fruits, such as the *Syzygium* heavy with red apples.

Walking along one day, one of Brierly's Aboriginal friends stopped to cut down a creeping vine and proceeded to make a rope to attach a *Pandanus*-mat sail to the mast of his canoe. In the mangroves, he pointed out the beautifully straight trees that made the best canoes. Other trees were used to make the floats of the outriggers. Brierly's guide continued through what could only be described as a plantation of sorts, agreeably bounded by a small stream,[47] to gather fruits that would be roasted later and tasted of chestnuts.

The Kaurareg greeted Brierly with excitement and delight on a return journey of the *Rattlesnake* in 1849, throwing their arms about him, offering food and giving him pride of place in a shaded spot.[48] The Kaurareg were massacred in 1869, by order of Police Magistrate Frank Jardine, ostensibly for the killing of the captain and crew of the *Sperwer*, at a place called Wednesday Spit— something in which the Kaurareg played no part.[49] With the aid of the Native Mounted Police and the crew of another vessel, Jardine's party advanced on the Kaurareg people, searching for the captain's wife and son, who, it turned out, were in Melbourne and had never been on the ship. An eyewitness account recalls the Kaurareg caught completely unawares. Huts were set alight, women and men screamed, guns were fired. Anyone fleeing was trapped and slaughtered. More reprisals were to come for survivors.

In 1872, a missionary traveller passed Muralag, where the Kaurareg were based: 'When we saw on Muralag the scene of the more than ample revenge exacted by the whites, we wished the Gospel had been introduced earlier to these dark places of the earth.'[50]

# CHAPTER TEN

# Evolution, interrupted

People made the land beautiful, but settlers took it because it was useful. Paddocks in forest gave them water, pasture, timber and security. By shaping land so carefully for grazing animals, people paved the way for pastoral occupation. The more carefully they made the land, the more likely settlers were to take it.[1]

Bill Gammage

Charles Darwin's theory of evolution by natural selection and survival of the fittest was never designed to apply to the social and economic affairs of men. But political cherry-picking of the theory, together with a further rattling of the Chain of Being, allowed for a searing shift that gave intellectual respectability to modern racism and justified views that certain races and classes of people were innately less fit and therefore more inferior than others. Dispossessing these people, usually native people, of their land— whether by force, collusion, ruse or violence—was merely following the laws of science.

The European invaders convinced themselves not only that it was the natural order of things to extirpate Aboriginal people, but also that it was necessary in order to unleash the flood of civilisation, shorthand for economic gain. Even the crudest ruffian, not tolerated in their own civilisation and looking for quick wealth in their search for gold—and hardly less nomadic than Aboriginal people—was a flag-bearer for civilisation and progress.

The inhumanity of the violent and often malicious conquest was easily justified by rejecting the very humanity of Aboriginal people. In Australia, Social Darwinism was particularly malevolent. Definitions of savagery that described godless anarchy, violence, cannibalism and sexual depravity meant that, by any rights divine or human, savages—Aboriginal people—forfeited their claims as owners of the land. Europeans were also relieved from any need to be kind and understanding. Killing Aboriginal people was considered less than murder because they were seen as less than human, so that mass murder became a way of life 'tolerated by the settler majority and winked at by the state'.[2] For many, murder and rape suited their vile carnal and cruel passions.[3] For the rest, there seemed no desire for a frontier policy that would mean restraint and concern for the Aboriginal people.

It would be an understatement to say that squatters were already predisposed to such views,[4] given the 500 or so massacres reported even before Queensland became a colony in 1859,[5] the same year Darwin published his soon to be perverted *On the Origin of Species*. There was no such thing as the noble savage, just the noble pioneer. This perverse philosophy is clearly on display in the note added by editor W.J.L. Wharton to Captain James Cook's journal of his journey in the *Endeavour*. Written in 1893, over a century after Cook penned his journal, Wharton's note recast Cook's views of a contented people so they suited the new cruel philosophical environment. Bluntly, and with breathtaking irony, Wharton aimed to correct Cook's view, accompanied by righteous fist-slamming capitalisation:

The native Australians may be happy in their condition, but they are without doubt among the lowest of mankind. Confirmed cannibals, they lose no opportunity of gratifying their love of human flesh. Mothers will kill and eat their own children, and the women again are often mercilessly illtreated by their lords and masters. There are no chiefs, and the land is divided into sections, occupied by families, who consider everything in their district as their own. Internecine war exists between the different tribes, which are very small. Their treachery, which is unsurpassed, is simply an outcome of their savage ideas, and in their eyes is a form of independence which resents any intrusion on THEIR land, THEIR wild animals, and THEIR rights generally. In their untutored state they therefore consider that any method of getting rid of the invader is proper. Both sexes, as Cook observed, are absolutely nude, and lead a wandering life, with no fixed abode, subsisting on roots, fruits, and such living things as they can catch. Nevertheless, although treated by the coarser order of colonists as wild beasts to be extirpated, those who have studied them have formed favourable opinions of their intelligence. The more savage side of their disposition being, however, so very apparent, it is not astonishing that, brought into contact with white settlers, who equally consider that they have a right to settle, the aborigines are rapidly disappearing.[6]

People were disappearing because Europeans were killing them. The mass killings and massacres were profligate, furtive and unprosecuted. And no perpetrator was ever legally punished. Indeed, how could they be when it was the 'police' who were the perpetrators? It is estimated that around 101,000 people were killed as the civilising frontier moved through Queensland[7]—a staggering number, and more than the total number of Australian deaths in World War I.

The indirect toll was much greater, with many more dying from disease, destitution and starvation.[8]

It is not as if the colonists didn't know what they were doing. With extraordinary frankness, the editor of the *Port Denison Times*, published in what is now Bowen, south of the Wet Tropics, noted: 'Thousands of our own aborigines have been shot down with calm indifference because they were troublesome on the runs, like kangaroos; they have been poisoned with strychnine, in company with dingoes; they have perished before our eyes from our loathsome diseases, and from our firewater.'[9] The *Cooktown Courier*, published north of the Wet Tropics, also chimed in that the way northern settlers were treating Aboriginal people was unparalleled globally in its horror. So much so that the editor dared not explain the exact manner in which the settlers dealt with Aboriginal people.[10]

———

On Thursday, 9 October 1873, George Augustus Frederick Elphinstone Dalrymple—the Father of North Queensland[11]—was in a boat on the North Johnstone River, the trunk that carries water from the drip tips of Thiaki to the Great Barrier Reef at today's Innisfail. Dalrymple was leading a government-funded expedition charged with exploring the last of Australia's uncolonised coastal land for its agricultural potential. Flanked by a steeply rising rampart-like escarpment to the west and the ocean to the east, the Wet Tropics had remained a barrier to Europeans a century after Cook had charted the coastline. After attempting some agricultural estimates, Dalrymple simply gave up and said it was inexhaustible.

Dalrymple was in raptures. A few days earlier, he had reported a magnificent red cedar, this one tree alone measuring nearly 7 metres in girth, at 1 metre from the ground. And now he described how the North Johnstone riverbanks clothed with dense masses of lofty forest, heavily festooned with flowering creepers of convolvuli, climbing bamboo, and lawyer palms, descend to the water's edge in steep slopes of luxuriant entanglement and variety of undergrowth; palms, bananas, ferns, lilies, arums, and large-leafed taro, struggling for prominence of position—a dazzling commingling of shades, colours, and intricate minutiae of outline . . . the deliciously scented arums, all in full bloom, and hanging in moon-flowers greeting us, as we passed, with whole greenhouses of rich perfume.[12]

Taro, he said, covered hundreds of acres in strips along the riverbanks as if they were being cultivated. Plantations of wild bananas of 'immense sizes and height' were also frequently found along the river.

A new adventure awaited around each bend of the sinuous river. On one particularly stunning site at the head of a north-west reach, he came upon rows of neatly made bark and palm-leaf huts of a large Aboriginal camp, noting that 'the blacks have certainly, as they frequently do, selected a most picturesque and commanding site for their head-quarters' camp: this one being placed at a sharp bend of the river, whence a beautiful view is obtained up and down two reaches'. These were Mamu people, and their huts were shaded by fine fig trees, eucalypts and *Castanospermum australe* (black bean) trees. Nearby was moored a small fleet of 20 catamarans, each capable of carrying several people. Dalrymple, seeing the camp empty, ascended to it and described a high, bare, flat-topped ridge of about half a hectare, swept perfectly clean and beaten hard by years—perhaps centuries—of dance and ceremonies.

At the front of his mind, however, was not the glorious wonder of the scene, or the mesmerising discovery that these people may have coaxed the rainforest into an even richer paradise by manipulating plants in a subtle system of domiculture.[13] His main concern was the inconvenience to his expedition of so many people. Known to exaggerate the aggressive and bloodthirsty nature of 'savages', Dalrymple observed the number of rafts on the river and said, with little evidence, that the Mamu were gathering for an attack. (They were more likely gathering for a ceremony, since the rainforest folk from as far afield as the Atherton Tablelands were known to travel to the coast at this time of the year.) But Dalrymple was itching for an excuse to invade the village so that the people could discover 'the full extent of their mistake'. Dalrymple's entourage of 26 included 13 Native Police working under Sub-Inspector Robert Arthur Johnstone (after whom Dalrymple would name the river); they were armed with new Snider breech-loading carbines, which could fire five times faster than the old muzzle loaders,[14] with large bullets that inflicted fearful injuries, literally tearing flesh apart as they flattened on impact.[15]

The next day, Dalrymple invaded the village, bringing up his entourage in three boats including one chartered by a speculator in quest of sugar lands. The police boat went ahead, with the people fleeing in panic, dropping dillybags with paints, nets, fishing lines and hooks, several packages of pounded root vegetable tied neatly in banana leaves, and a basket full of human bones (carried as solemn mortuary practice). Johnstone dispersed them. In the afternoon, after establishing camp, Johnstone 'again sent the troopers out, and went and found the blacks closing in on the camp and dispersed them; and during my watch at night, they attempted to stalk the camp, but were frustrated; and in the morning they came below the camp and challenged us to come to them. We dispersed them.'[16]

'Dispersal' was a euphemism for shooting to kill. It was a well-known colonial cold joke.[17] The fact was that the Native Police comprised the most efficient and well-trained killing machines in the world, especially at their peak during the 1860s, 1870s and 1880s. It wasn't the first time that Johnstone had taken deadly aim with his Snider along the river that was to be named after him.

---

During the year before, 1872, a group of gold-seekers—young Sydney bloods, many from prominent families including the son of former NSW Premier William Forster—set off for New Guinea in the *Maria*, an old, leaky and unseaworthy coal barge. It was a boys' own adventure pursuing a mythical pot of gold.[18] Ill-equipped, ill-prepared and with an inexperienced captain, they were shipwrecked on Bramble Reef off Rockingham Bay on 26 February. While several men drowned, most escaped in three boats and two rafts. Two of the boats, with 28 men, made it to the newly established Cardwell on 3 March. Two men from the third boat, which was left beached near where the Kennedy Expedition had come ashore, stumbled nearly dead into Cardwell on 6 March. Apparently, they had been attacked by Rockingham Bay people, and the captain and others from that boat were assumed dead.

A party, presumably including Johnstone, was sent to find the boat, and they spied it drawn up into coastal forest. When they hauled it into the water, they were attacked; eight rainforest people were killed in the ensuing fight. The rescued boat was rowed to Cardwell and, being in good condition, was renovated to become Johnstone's police boat. In the meantime, Police Magistrate Sheridan in Cardwell requisitioned the newly arrived HMS *Basilisk*, returned from New Guinea after carrying out hydrographic surveys, to assist with further action against the rainforest people whom Sheridan exaggerated would otherwise attack boats and Cardwell itself. Captain John Moresby reluctantly agreed to transport Johnstone and the Native Police. Moresby later reported that the punishment meted out to the people was severe and ugly, and it was done with an unrestrained ferocity that disgusted him and the officers of the *Basilisk*.[19] The same people who sat politely at the line in the sand in 1848 were no more.

The two rafts of the shipwrecked *Maria* had meanwhile landed: one north and one south of a large river newly discovered by the *Basilisk* and named the Gladys by Moresby. (The following year, Dalrymple would conveniently ignore this and name it

177

the Johnstone.) The *Basilisk* continued searching and found eight bedraggled survivors from the northern raft. They had been taken in by people and fed, initially, on fruits and nuts. The survivors were then helpfully rafted across a tributary to join another group of people who had caught a haul of fish. Fires were lit for the comfort of the survivors, who mostly slept in huts including one freshly made for them and strewn with grass. Here, they dined on bananas, yams and roasted cockles. The next day, they were led to a place where another family was cooking yams, bananas, fern roots, crabs and prawns: it seemed that the people were moving the survivors around to share the load of feeding their voracious appetites.

On the morning of 12 March, the people accompanying the survivors suddenly began shouting and gesticulating wildly while pointing towards a vessel. There, in the distance, was the *Basilisk* still scouring the coast for survivors. Spying the survivors, Moresby himself came with the ship's boat. He was quite affected when the people embraced the survivors and wept as the boat pulled away with them.

The fate of those on the second raft was not to be the same. The *Basilisk* found the second raft, together with the bodies of three men. A steamer, HMS *Governor Blackall*, sent by the government to assist in the search for survivors, was also nearby. It had on board Johnstone and his Native Police, who claimed that some of the shipwrecked sailors had been eaten and others very recently murdered. It is possible that the survivors' deaths were retribution for the massacre that had occurred just the week before when the people saw what had become of their kinsmen at Johnstone's hand.

In a crescendo of retributive zeal, an army of vigilantes was placed under the command of Johnstone and his troop. Five boat-loads of Cardwell residents conducted reprisals for 50 kilometres up the coast, including in the Country of the people who nurtured the eight survivors of the first raft. Follow-up raids continued until 1878: one of the most populated places on the coast was reduced to remnants by 1886.[20]

———

Back on the newly renamed Johnstone River in 1873, Dalrymple continued his explorations. Much to his annoyance, another group of Aboriginal people came along the river. Once again, he ordered the Native Police to deal with the locals while he, undeterred, set off up the river for some exploration. The river soon shoaled, bringing that day's exploration to an end and fixing the river's navigable capabilities. The next day, being Sunday, Dalrymple's party held a religious service 'read under the pleasant shade of the large trees on this old scene of cannibalism and savage rites'.[21] Without a backward glance, the party returned to the coast the next day to continue their planning for clearing the land of its forests and peoples in order to make room for a major port development, canals, wharfs and railway lines for sugar cane.

Dalrymple and his crew then continued to sail north, and he waxed lyrical in a grand description of the stunning coastline, now mostly the Wet Tropics of Queensland World Heritage Area. They sailed past the Seymour Range, with its manicured woodlands alternating with open glades and grassy ridges. Dalrymple named a northern extension of this range after the secretary for public lands, Charles Graham. He spied the joint opening of the Mulgrave and Russell Rivers, which, on the return journey, he named and explored, noting the frequent appearance of outrigger canoes and numerous round-topped palm-leaf huts on the banks at the bends of the Mulgrave River, clearly an industrious and settled country.

Coming into view, as he sailed north, was the Malbon Thompson Range 'all clothed in dense dark green jungles', from which rose two lofty peaks: Bell Peak North and Bell Peak South. Recent biological surveys have revealed that these peaks harbour endemic plants including a rhododendron, and beetles that are relict from past wet climates. Dalrymple marvelled at the great white flocks of Torres Strait pigeons that flew from the jungled slopes and across the sea in the evening to roost among the trees of the sparkling offshore islands. He noted a school of about 20 large turtles passing in a line towards the shore, 'ploughing through the sea with considerable power and velocity, their great square heads throwing the water aside'.[22]

On Fitzroy Island—Gububarra or Song-giver[23] of the rainforest people near Cairns—Dalrymple bemoaned the fact that vandals from some vessels were clearing the shady figs and magnificent *Calophyllum* trees. With unconscious irony, he complained that without 'legislative enactment' many places along the coast would be despoiled.

Spying an unknown river opening on the southern edge of Trinity Bay, Dalrymple took his crew of Native Police and sailed on the ship's tender for what would be the site of today's Cairns. He named a perfect pyramidal mountain Walsh's Pyramid—known as Djarragun (the hill of the scrub hen's nest) by the Gulgibara Yindinyji—after his friend, William Henry Walsh, who would soon be the Speaker of the Queensland Legislative Assembly. The indigo-coloured bulk of Bellenden Ker rose into the clouds behind Djarragun. Together with Mount Bartle Frere, the Bellenden Ker range is one of the most significant ranges in the Wet Tropics, indeed Australia. With its rich diversity of habitats and its height, it acts as an evolutionary time capsule for ancient and disjunct plants and animals including endemic frogs, snails and spiders, and bugs known only from the very top of the mountain where they cling little changed from Gondwanan times. Dalrymple mused that no European had penetrated these mountains.

Sailing north, his party anchored at Double Island, offshore from today's Palm Cove. Here, a shore party led by Johnstone and his troop opened their Sniders on a group of Yirrganydji warriors who, with understandable alarm, were ready to defend their Country against the impolite intruders. Dalrymple later crowed 'from the discoveries made in their camps we all heartily rejoiced at the severe lesson which their unwarrantable hostility had brought them, and that such blood-thirsty bullying scoundrels had at length met more than their match'.[24] With gleeful certainty, Dalrymple noted evidence of wholesale cannibalism: roasted and partly eaten bodies, and lumps of half-eaten human flesh found in dillybags. The meat and long bones observed were more likely from a cassowary.

After this massacre, his men climbed Macalister Range, which Dalrymple named after Arthur Macalister, a premier of Queensland. They noticed a jumble of low ranges further inland. These wooded ranges, out of which looms Black Mountain—situated in between the rainforested highlands to the north and south—are cloaked not in wet rainforest but in a matrix of dry rainforest and savanna. This is known as the Black Mountain Corridor. It is a dry corridor that, at times, forms a biogeographic barrier to movement of species. Climate oscillations connect and disconnect rainforests in a chaotic pattern across this corridor so that species are repeatedly isolated and joined in what is a forceful species pump. In the current Holocene interglacial period, rapid hybridisation is occurring as species—such as the northern bettong, the chowchilla and a suite of reptiles, frogs and orchids separated during the last glacial period—once again meet across the wild confusion of valleys and watersheds observed by Dalrymple's men.

Further north, Dalrymple noted the grandeur of the mountain background of the northern highlands of the Wet Tropics of Queensland World Heritage Area, naming them the Heights of Dagmar and the Heights of Alexandra, presumably after Danish princesses who were in the news at the time. He passed the opening of a river that he would later explore and name the Daintree after his friend, Richard Daintree, the agent-general of Queensland.

Dalrymple recorded that red cedar was plentiful here. This was to cause a cedar rush within months. Behind Dagmar and Alexandra are the 'lofty granite peaks of most picturesque outline' of Thornton Peak. One of Queensland's highest peaks, the Thornton uplands house a suite of endemic plants and animals, including some snails and a water beetle that have evolved to live in the cool litter of the mountain tops. The boulder fields skirting the mountain are strongholds for the ghost bat. Unique and limited plant communities thrive around the spring-fed creeks seeping from the mountains.[25]

Reaching the Endeavour River on 24 October 1873, Dalrymple's party camped on the same quiet beach as Cook had a century before. The very next morning, they were startled by the appearance of tall masts and yards appearing over the mangrove belt that soon resolved into a large private steamer, the *Leichhardt*. The steamer had a complete government staff of police to support a gold commissioner. And there was an engineer who had come to make a road to the newly discovered Palmer River goldfield.

Dalrymple mused that just the day before they 'had sailed into a silent, lonely, distant river mouth, with thoughts going back a century to the arrival of the brave navigator . . . in knee breeches, three-cornered hats, and small swords, pigtails, and silver shoe-buckles'. The next day, they were in the middle of frenzied activity 'characteristic of the present day—of a young diggings' township— men hurrying to and fro, tents rising in all directions, horses grazing and neighing for their mates, all around us—the shouts of sailors and laborers landing more horses and cargo, combined with the rattling of the donkey engine, cranes and chains'.[26]

In 24 hours, the pincer movement that would raze much of Thiaki's world was complete.

# CHAPTER ELEVEN

# The other side of the pincer

The Wet Tropics is long and narrow, perhaps 70 kilometres at its widest point. Flanked by the Great Barrier Reef on one side, on the other are the vast savannas where squatters aided by the Native Police closed in on the Wet Tropics from the west. Squatters, in the flowery classical allusions of the time, 'went further than the man before him. There was plenty of grass and travelling was an honourable and recognised quantity in those Arcadian times when pastoralism was a quest in harmony with nature.'[1] Nothing could have been further from the truth.

Maritime explorers, by and large, did not come from a perspective of treating people as if they were to be conquered and dispossessed of their lands. Squatters, however, had precisely that aim. Any good intentions of the first British governors—who were, in British law at least, the ultimate defenders of Indigenous peoples[2]—evaporated with the spread of squatters, as the lure of new country took men further and further into lands that were out of government control. Governor John Hunter (1795–1800) allowed settlers wide power to fight Aboriginal resistance, and Governors Philip Gidley King (1800–6) and Lachlan Macquarie (1810–21) vigorously used military force. Governor Thomas Brisbane (1821–5) created martial law in the region around today's Bathurst in New South Wales, effectively legalising the slaughter of the Wiradjuri under the protection of it. Governor Ralph Darling (1825–31) urged settlers to combine to defend themselves, to 'pacify' Aboriginal people and to repel those who approached their land, village or farm, allowing settlers to take whatever action was required.

Captain Cook's view had set the tone. Because Aboriginal people were, according to Cook, few and timid, Australia was treated as a colony of settlement and not of conquest. Aboriginal people were, strictly speaking, British subjects. Therefore, any rights based on prior occupation could be ignored, and Aboriginal people could be driven off the land. In 1836, this was mandated in a NSW court (*R v Murrell and Bummaree*);[3] the ruling legally dispossessed Aboriginal people of their land and stated that they did not have land rights. Perversely, this meant that squatters could demand the protection of British law for 'self-defence' actions against Aboriginal people who, in defending their ancestral land, were effectively criminals.[4]

By 1836, it was all a moot point anyway. Violence against Aboriginal people had long before shifted from self-defence to punitive and pre-emptive, as armed Europeans invaded populated landscapes aided by unsupervised convict and ex-convict labour. British law rushed to catch up with the squatter by dividing the land into districts and charging a small fee for the Border Police, usually ex-convicts, to sweep away any Aboriginal resistance.

In 1838, Governor George Gipps (1838–46) tried to enforce the rule of law impartially on both white and black. A massacre of Aboriginal men, women and children at Myall Creek on the New England Tablelands in New South Wales saw the execution of seven cattle men. This, however, merely taught the invading squatters to be more circumspect. They, instead, cleaned up more carefully after massacres, or resorted to less gruesome ways of killing, such as poisoning. Despite Gipps' efforts, the 1840s ushered in the worst racial violence in the colony's violent history. In desperation, Gipps formed the Native Police in New South Wales, which subjected young Aboriginal men to military discipline and used them 'to preserve the peace'. It was disbanded in 1842 but a new force was set up under Governor Charles FitzRoy in 1848 and the experience was passed on to Queensland.[5] Established in 1849, the Queensland Native Mounted Police was a government-financed paramilitary force tasked with protecting the advancing squatters and subduing Aboriginal resistance.[6] It became one of the most violent forces in Australia's history, particularly in Queensland.

In the newly established Queensland, the administration set up a nakedly pro-squatter select committee in 1860 to review the NSW lessons of frontier policy. With the security of the frontier a major concern, Aboriginal people were cast as cannibals that had sunk to the lowest depths of barbarism. The select committee dismissed them as beyond civilising; they were a doomed race already dying out. And so, they were set upon by the superior mobility and organisation of the Native Mounted Police, who had European weapons and powerful allies among the squatters. Aboriginal people didn't stand a chance.[7]

Why would Aboriginal men inflict such violence on Aboriginal people?[8] Information is difficult to find. Traumatised, destitute and unmoored from their traditional lives, devastated by the violent treatment of their mothers, wives and daughters, and utterly unable to participate in white society on equal terms, some young men must have chosen at least a small degree of empowerment. Coercion, kidnapping and inducement were also common, since desertion was relatively frequent even though, if caught, it was accompanied by extra-judicial executions and floggings close to the point of death with whips or wire. There were also attempts at suicide.[9]

The employment of Aboriginal men from distant regions ensured no kin affiliations: the idea of pan-Aboriginal solidarity is largely a contemporary concept. A trooper's presence in foreign territory actually made it dangerous for them to desert. Troopers were also carefully selected from 'semi-civilised' tribes who may have been educated in missions or those known for bushcraft, some of whom could 'track a mosquito over a stone wall'.[10]

The Aboriginal side of the Australian Wars is rarely seen. One Maranoa corroboree was recorded in the early 1860s, at a time when cattle runs were being taken up with astonishing rapidity, to celebrate a string of victories against the European invaders. The corroboree began with a group of actors representing a herd of cattle: 'Some lay on the ground lugubriously chewing the cud; some tried to scratch themselves with their hind feet, some stood licking the young calves or rubbing their heads together.' A group of warriors then darted into the group of snorting cattle, spearing a couple while the others stampeded off. The fallen cattle were skinned, cut up and swiftly carted away. A subsequent scene depicted a group of white men on horses:

> White ochre smeared over aboriginal faces gave them a ghostly hue under their imitation cabbage-tree hats. Bodies were painted blue and red, the prevailing colours of white men's Crimean shirts. Swamp reeds bound round the legs were a fair representation of the cowhide leggings which the settlers wore as protection against the brigalow scrub. The horsemen wheeled to the right, unslung their carbines, and went through all the motions of loading before firing a volley into the fleeing body of a blackfellow before them. Occasionally a black man went down, accompanied by groans from the audience, but it turned out to be only a trap. Suddenly the whites were confronted by natives coming from everywhere and turning their horses' heads they galloped furiously for safety. The audience by now frantic with delight cheered uproariously as the last white man stumbled over a root and disappeared from view in a cloud of dust.[11]

---

When the colony of Queensland was proclaimed in 1859, only the south-east corner was occupied. Kennedy's travails of 1848 stalled the movement of the squatters to the north for a short time. But Augustus Charles Gregory's expedition of 1855–6 laid open, like an incision, a line across the north from the Elsey, Roper and Macarthur Rivers down through Queensland by way of the Flinders, Burdekin and Fitzroy Rivers. Pastoral runs were established at Gracemere, on the Fitzroy River, in 1855; nearby Rockhampton, the northern-most town at the time, was declared a port in 1857.

It was from here that Dalrymple led the first of his expeditions to the unexplored lands north of the Tropic of Capricorn. While some pastoral districts had as few as one Native Mounted Police camp, the North and South Kennedy Pastoral District—a vast area from Cardwell to Mackay, west to the Great Dividing Range and the entire Burdekin River system—had 24 of them. This was a gruesome reflection of the speed at which Aboriginal presence was controlled or exterminated: it was easier to run down families of Aboriginal people in the open plains of western Queensland than in the rugged jungle-clad topography of the Wet Tropics. The fact that most Native Police camps were established for less than a decade also exposes the horrific speed of the effacement of Aboriginal people.[12]

Each Native Police troop consisted of a senior European 'sub-inspector' who drilled a group of four to ten Aboriginal troopers. They patrolled pastoral stations, moving 20–30 kilometres a day, 'giving any troublesome blacks an occasional lesson'.[13] In a parody of justice, complaints of squatters were generally accepted on face value. But punishment was collective: the whole tribe was massa-cred except perhaps for women, who became 'camp followers'. On patrol, the troop had unchecked power that assumed the roles of police, counsel, judge, jury and executioner.

Dalrymple relied on the Native Police. In 1859, he formed a syndicate to explore the area north of Rockhampton to the Burdekin River and its Valley of Lagoons, reported by Leichhardt in 1845 to be fine pasturing country where water, grass, hills, mountains, plains and forest were all united.[14] When he returned, he found that Queensland had been proclaimed a separate colony. And while his own claim on the Valley of Lagoons was temporarily rejected as speculative, he was compensated with an appointment as the commissioner for Crown lands in the Kennedy District.

With this under his belt, he set out in the *Spitfire* in 1860 to find a port for the produce of the Kennedy District and his future pastoral estates in the Valley of Lagoons. The mouth of the Burdekin River proved useless for navigation. Instead, he established a settlement at the newly discovered Port Denison (now Bowen), about 80 kilometres to the south. The area was densely populated, and Dalrymple—who was always quick to use firearms against any Aboriginal resistance—took charge of the takeover by advancing in a line with Native Mounted Police. He subsequently recommended a chain of police outposts to allow safe white settlement.[15]

A year later, a boat with officials, settlers and stores sailed from Rockhampton to Port Denison. Dalrymple, accompanied by Native Police, led a forward land party and 'frightened off' people still living on Country. Unlike other pastoral areas where the invasion took the form of a more gradual infiltration, albeit with the same murderous outcome, the occupation of Port Denison was a rushed invasion of menacing proportion. The local people were invaded suddenly by sea and land, the Native Police being used as an advancing army.

Bowen's foundations thus cemented in blood, the Native Police subsequently scoured the river valleys of the vast Burdekin catchment. People fought back with a ferocity that became legend, but within six weeks pastoral runs had been taken up along the Burdekin and inland, and by 1863 the whole of the Kennedy District (apart from the Wet Tropics)—an area larger than England—had been taken up with stock 'to fill the land everywhere with the beginning of civilization'.[16] Dalrymple took up his large pastoral runs in the Valley of Lagoons with wealthy friends—including the first premier of Queensland, Robert Herbert, as a silent partner—in 1863. In 1869, only 30 Aboriginal men survived in the now renamed Bowen, when just a few years before there had been hundreds of men. An eyewitness commented that when the Native Police visited the public house after their ravages around Bowen, the heels of their boots were covered with brains, blood and hair.[17]

Dalrymple needed a port closer than Bowen for the produce coming from the Valley of Lagoons, and he set upon Cardwell, which was already well surveyed. He chartered the schooner *Policeman* in 1864 and loaded it with owners of other neighbouring estates, several Native Police, other squatters ready to take up land, bushmen, a botanist, a publican, a storeman, a carpenter and a market gardener—a veritable pop-up town. They anchored 1600 metres offshore and celebrated their arrival as pioneers of Cardwell with a fanfare of fired guns and rockets (making some deaf for a week[18]).

Dalrymple had with him James Morrill, a shipwreck survivor, who had the year before found his way back to 'civilization' after spending 17 years from 1846 to 1863 with Aboriginal people, relatives of the Bowen people. While his people's Country was centred further south from Cardwell, he could still make himself understood. When canoes of local people came to *Policeman*, he warned them off.

The next morning, with horses, sheep, goats, fowls and dogs landing, the local men approached again, wanting to know if the Europeans came as friends or enemies. Morrill, at Dalrymple's direction, said they came as friends, but he again warned that they must clear out and also tell others to do the same, as the land was going to be occupied. He emphasised that the Europeans were strong, and more would come. With looks of disdain, the Girramay men told the Europeans to leave, and proudly walked away.[19] Three days later, Dalrymple and his Native Police came upon a party of armed Aboriginal men. They were set upon and left 'rather cut up'.

———

Once Dalrymple had secured the site of Cardwell, he began invading along established Aboriginal paths from Rockingham Bay west to the Valley of Lagoons. He passed through a dense and lofty forest and after a few kilometres 'entered a very beautiful tract of rich country' with 'openly timbered ridges descending from a range into small rich plains and forest glades, intersected with many clear running stony streams'. From an amphitheatre of precipitous jungle-clad mountains, a river cascaded into the woodlands. 'A broad, hard-beaten path of the blacks led . . . into this retreat, where small verdant plains, bounded and broken by clumps of vine, jungle, and fig-trees, [were] varied by the fresh, bright green of groves or single trees of wild banana', as well as the tall and graceful stems of the edible *Ptychosperma elegans* (solitaire palm), endemic to Queensland and now often cultivated in gardens. These 'half completed the delusion that we were entering one of the beautiful mountain villages of Ceylon'.[20] The people had minutely managed Country for variety. Integrated across Nations to fit the universal fabric of Dreaming, it was a beautiful and extraordinary paradise.

Where Dalrymple ascended the ranges, he found 'a line of perfectly open, bald, grassy summits for about two miles'. These descended steeply to the plains of the Herbert River, winding

like a silver snake out of the gorges of its upland birthplace, through mountain-flanked rich woodlands and plains. Further to the eastward it spreads out into the dim distant level seaboard of Halifax Bay, with its faint blue line of ocean, dotted with the hilly outlines of the Palm Island far to the seaward; all softened and mellowed by the gauze-like summer-heat haze of declining day—the setting sunbeams shining deep purple on the distant crenelated peaks of Hinchinbrook, and the chains of mountains north and south. Most grand and lovely in its scenery is this 'Vale of Herbert'—mountains, peaks, cliffs, waterfalls, forests, plains, and what is seldom met with in Australian scenery, the clear waters of a broad running river, adding life, light, and beauty to the whole.[21]

———

An account of Morrill's life, *17 Years Wandering Among the Aboriginals*, was co-written by E.B. Kennedy,[22] who had served as an officer in the Native Police and had just returned from patrolling the Country of Morrill's kinsmen, the Biri.[23] Kennedy's interests may have been predatory, and the nature of his questions—are they cannibals, do they practise polygamy, do they have any religion—seemed intent on measuring their savagery in order to justify bringing on their demise.[24]

Morrill's voice, however, speaks clearly of loss of kinsmen and family. He mourned for a friend shot dead by a sailor while trying to alert the sailor to the fact that a white man was in their tribe. It was 1860, and the sailor was from the *Spitfire*, which was carrying out Dalrymple's Burdekin River explorations. Morrill's Aboriginal family reported 'a lot of white and black men on horseback, near Cape Upstart, shooting down the tribe'.[25] This presumably was the Native Police with squatters and their cattle. Morrill's kinsfolk barely believed that the squatters ate these strange beasts with big eyes and long tails, and rode the horses with saddles and stirrups. The Biri found it incredulous that these apparitions had come to take their land. But reports filtered in daily of cattle in great numbers and apparitions on horseback with a stockwhip that sent the people fleeing up into trees in fright. The cattle drank all the water in the waterholes before the Biri could even get the fish out.

Morrill then heard that around 15 of his kinsmen had been shot dead while fishing. He convinced his people to show him where the whites were, explaining that he might be able to save people's lives. The tribe travelled to a hunting ground and, while they spread their nets to catch wallabies, some old women were sent as spies to look for the whites. They brought word back that the whites lived in a large hut, with blankets hanging on the stockyard fences, a dog barking and a sheep tied up to a tree, bleating. Morrill decided to go to them. Tribal Elders would not let him go alone, but his companion fled in fright at the first sight of the grazing sheep. Morrill found a waterhole to scrub himself as white as possible. He carefully walked to the hut, not knowing what to do, and got on the fence to prevent being bitten by the dogs. He called out, 'What cheer, shipmates?' A white man came out and then called back to his mate in the hut to bring the gun because there was a red man sitting on the fence. Morrill called, with rusty English, 'Do not shoot me, I am a British object—.'

After close questioning, the invaders gave Morrill bread on which he choked. He was not hungry anyway, having feasted on wallaby that day. The squatters offered Morrill clothing, but he said that he needed to go back to his people to explain the situation to them. The squatters replied matter-of-factly that if he didn't return the next day, they would put the native trackers onto him and his tribe.

Morrill returned and was taken first to Bowen and then to Brisbane, where he was publicly baptised. He came back to Bowen, where he died two years later: his symptoms of hard swellings and pain suggest the final stages of skin cancer. Through his accounts, Morrill pleaded for the safekeeping of his tribe, who had been so kind to him, saying that 'almost their last wish to me was with tears in their eyes that I would ask the white men to let them have *some* of their own ground to live on. They agreed to give up all on the south of the Burdekin River, but asked that they might be allowed to retain that on the other, at all events that which was no good to anybody but them, the low swampy grounds near the sea coast.'[26] Here there were roots and yams, and fish in the rivers. The plea landed on deaf ears, as Morrill must have known it would.

# CHAPTER TWELVE

# Desultory little massacres[1]

Just months after Dalrymple completed his 1873 survey along the Wet Tropics coastline, cedar-getters were already cutting timber in the forests of the Daintree. One party had cut down 70,000 super feet of logs and rafted them down the Daintree River to Cooktown.[2] So greedy was the rush that, by 1877, the red cedar of the Daintree and the adjacent Mossman valleys was virtually cut out, and there were reports of starving rainforest peoples as cedar-getters invaded the rich slopes and streams of the rainforests.

Deeper in the forests, a quiet yet determined war was taking place as people fought cedar-getters and miners, who subsequently disappeared, presumably killed. Unlike in earlier decades of the nineteenth century, people now showed little mercy to any shipwrecked survivors. A record exists of a party of eight seaborne cedar-getters who were attacked at the Johnstone River mouth in 1877; their goods and boat were taken and burned. They hurriedly made a canoe out of a cedar log and drifted north, washing up at Port Douglas five days later from where they returned 'better equipped'[3] to deal with the people.

In 1878, it was reported in the local newspaper that over 4 million super feet of red cedar had been taken from the Daintree and Mossman valleys on 59 ships. Another million awaited shipment from the Mulgrave River. These estimates were known to be conservative.[4] By 1887, there were 59 timber licences issued in Cairns; by 1890, millions of super feet of red cedar had been transported from Clump Point north of Mission Beach,[5] halfway between Cardwell and Innisfail.

One historian reported that when settlers took land in the Johnstone River catchment for sugar, they were mostly unmolested because the cedar-getters had dealt with the Aboriginal people.[6] Robert Arthur Johnstone was still on the scene. In 1880, he tracked the killers of a worker from a surveyor's camp near Victory Creek where he had, seven years earlier, invaded the rainforest village. Stripped down to hat, loincloth, cartridge belt, rifle and boots for stealth, he set out with his troopers 'who were not so modest'[7] to track the group of supposed murderers. He hunted and killed all bar one, who was left to warn any others.

———

Further south, on the Herbert River mouth, 20 kilometres southeast of Cardwell, Carl Lumholtz had observed in 1882 the frenzied effort to clear the forest for sugar cane:

> The large flocks of pigeons had difficulty in finding the high quandong-trees, in which they are wont to light. The magnificent 'weaver-birds' [metallic starlings] flew about homeless in large flocks, for the great trees in which the colony had their numerous nests were felled. The cassowary became more and more rare; still, I could see its footprints in the sand.[8]

Lumholtz, a Norwegian explorer and ethnographer, was one of the few Europeans outwardly critical of the treatment of Aboriginal people. 'The Government of Queensland annually distributes blankets to the natives on the Queen's birthday,' he said, 'if they will but come and get them. This is the only thing the Government does for the black inhabitants.'[9] Lumholtz spent a year with the upper Herbert River people who still eked out relatively traditional lives in mountainous strongholds, 100 kilometres from the coast. They provided a rare glimpse of the daily lives of rainforest peoples.

Lumholtz was often aghast at the uncivilised brutality of the whites. Yet he remained shackled to the ideology of the Chain of Being. On the one hand, as if by rote, he described people as among the lowest humans to be found, a doomed race and, inevitably, cannibals. Then, in complete contradiction, he admired their pose, their skill, their presence. They carried themselves, he said, 'as if conscious that they are the lords of creation'. He was charmed by the people who were accompanying him in his quest for new species. Lying on their backs, they sang unselfconsciously in the evenings while accompanied by clap sticks: 'A black man rambling among the trees alone may at times be heard making the woods echo with his joyful song . . . free and happy in his native hunting ground.'[10]

He watched as the local people climbed trees as easily as Europeans walk upstairs. He was amazed at how efficiently a hut was made; how crayfish were caught by making a low babbling sound and then spearing them; and how spears, nets and baskets were crafted, particularly the honey baskets so closely joined that they held water. He was enthralled with the intense observational power of the people. Nothing escaped their notice as they carefully examined orchids and ferns for rats and pouched mice; searched the fallen leaves for *Hypsiprymnodon* rat-kangaroos; and took frequent handfuls of dirt or litter out of a crevice in a rock, or a cleft in a tree, and breathed in the scent to see if any animal had recently passed. 'They sought out every trace to be found, took notice of broken branches and bark, or of stones that were turned, or of a little moss that had been rubbed off: in short of everything that would escape a white man's attention . . .'[11] In this way, people would detect any strangers, black or white.

He observed that family groups across a tribe occupying an area he estimated to be around 3000 square kilometres were on good terms and had wide individual liberties if Country and Law were respected. His companions would ask him to shoot the land so that it would frighten and alert strange tribes.

The rainforest people revelled in existence and reacted to pleasure and pain with immediacy. Lumholtz loved their keen sense of humour and drollery. Their ability to mimic was extraordinary and produced howls of delighted laughter when white men were the brunt of the hilarity. Lumholtz recalled an incident while rambling about in the woods. As usual, the people were paying close attention to the bees, and often gazing up in the trees. He could never match their phenomenal eyesight. But one day he did: he was the first of the group to observe a small swarm of bees a couple of metres up from the ground. The group was greatly astonished that a white man could find honey, and one man expressed his feelings by rolling in the grass with mirth.

The people helped Lumholtz 'discover' four new mammals: Lumholtz's tree-kangaroo—the most beautiful mammal he'd ever seen—and three possums (the green ringtail possum, Herbert River ringtail possum and lemuroid ringtail possum). He also collected 700 specimens of birds and other animals.[12]

While Lumholtz frequently commented on the fertile and productive patchwork of grass, forest and rainforest scrub, he missed entirely that it was the handiwork of the people. In this he was no different to most ethnographers. The anthropological literature is scant indeed about Aboriginal land management. Of the 3600 published records made before 1970 of Aboriginal people in the savanna, for instance, only six per cent referred to any form of applied land management. And then only superficially.[13]

———

So dismissive of the rainforest peoples were the first Europeans to visit the Atherton Tablelands that very few details at all were recorded about them. Apart from Kennedy, the first Europeans in the region were cedar-cutters and miners, and they were mostly not inclined or unable to record details of the new environment and Aboriginal customs and land management they saw. Miners documented only a conical hill holding a crater at the head of the Barron River—probably Mount Quincan—and a waterfall on the Johnstone River.[14]

The documented history of European settlement can be traced back to prospector James Venture Mulligan, who reported rich alluvial gold along the Palmer River in August 1873—setting off the gold rush that Dalrymple witnessed in Cooktown in 1873.[15] By 1877, Cooktown boasted nearly 100 hotel licences catering for the 7000 European miners rushing through to the goldfields in northern Queensland. These were followed by 17,000 Chinese miners, who more patiently worked the alluvial pickings once the frenzied digging was done.

By 1880, most miners had moved to the Hodgkinson goldfield, also discovered by Mulligan (in 1876), in the rough and inhospitable ranges on the western edge of the rainforests visible from Thiaki as a smudge on the north-west horizon. Here, there were reported to be 20,000 miners in the field. The pioneer communities springing up around the mines were not big on maintaining strong moral leadership—there were 1000 licensed premises, one for every 36 Europeans.[16] The ore was transported with difficulty over a back-breaking track across the precipitous coastal range and down to the coast, resulting in the establishment of Cairns in 1876.

The goldfields were on or near watercourses essential to Aboriginal people, blocking access to food. Starving people were compelled to eat horses and bullocks, which they mustered into rugged strongholds from where they also attacked small groups of miners and teamsters.[17] Aboriginal warriors logically had no choice. There was no value in not fighting since the result either way was just the same—death.

Between the discoveries of the two goldfields, Mulligan continued to prospect, moving closer to Thiaki. In April 1875, he picked
his way up the Barron River, across the junction of Emerald Creek
and Granite Creek—the site of today's Mareeba—and rode on
southward over rich basaltic soils. Near the site of Tolga, he found
a wall of dark impenetrable jungle and turned westward to skirt
around it, marvelling at the giant red-cedar trees. On 3 June, he
found a well-beaten track that was frequented by Aboriginal travellers wandering between the western perimeter of the rainforest and
the Mulgrave River on the coast via forest pockets:

> [T]he native track passed on the west side of the scrub
> right between the range of hills and the scrub. A splendid
> track, the best native track I ever saw anywhere. There are
> roads off the main track to each of their townships, which
> consist of well thatched gunyahs, big enough to hold five or
> six darkies. We counted eleven townships since we came to
> the edge of the scrub, and we have only travelled four miles
> along it. At certain seasons this must be a crowded place
> with blacks, which seem to live principally on nuts, for there
> are barrowsful of nutshells at their camps . . . The natives,
> I think abandon this country at this long grass, wet season of
> the year. Their paths are well trodden, and we follow them
> sometime for miles.[18]

The people were likely visiting the saltwater creeks and coastal areas
at the time Johnstone was carrying out his retributive raids for the
murders of the *Maria* shipwreck survivors in 1872.

The next day, Mulligan's party passed over the shield volcano on
which Atherton now sits and, getting a clear view, began to ascend
a rough granite range, the Herberton Range, the western flank of
Thiaki. On the Wild River, Mulligan discovered tin on 5 June. He
thought that it wasn't of much value, considering the cost of transport to Cooktown.[19]

In 1877, Christie Palmerston—after working on the Palmer and Hodgkinson goldfields—had located an easier route to the coast at Port Douglas. Palmerston was one of the most famous bushmen and prospectors of northern Queensland, and regarded as the Prince of Pathfinders.[20] His easier track to Port Douglas had made transport viable, and the tin town of Herberton sprang up in 1880 along the Wild River of the Barbaram people.

A miner was killed invading Barbaram territory, and a detachment of Native Police was established to protect the miners who descended on Herberton in droves, resulting in 'the almost instant elimination' of the entire Barbaram people.[21] It soon became sport for Europeans to shoot on sight any people found lurking. One Herberton resident boasted of killing 15 people in a day.[22]

———

The establishment of Herberton also triggered a rush of another kind: the final explosive red-cedar rush of 1881. With the entire east coast having been logged out, this was the last domain of red cedar in Australia—and Thiaki was in the blast zone. In Brisbane, William Alcock Tully, under-secretary for public lands, had already warned parliament that north Queensland was a timber quarry. Trees were being removed faster than they could be replaced, for which the government was receiving nominal fees from a few timber licences. Tully's complaints were ignored. And much of the red cedar was hacked down illegally in a frenzied attack. So wasteful was the felling that rare and priceless timber became the common building material used for house frames, weatherboards and dunnies. It was chipped for roofing shingles and chook houses, and used down mine shafts.

In 1882, cedar-getters were felling huge giants that yielded as much as 25,000 super feet from the barrel alone.[23] Common practice held that sawyers could claim no more trees than they could cut at a single pit, but this was thrown to the wind with the sawdust as sawyers cut down all likely trees, including young trees, as quickly as possible, lest some other sawyers got there first. The best logs were pit-sawn by crosscut saw into boards, and the rest abandoned to rot. While unlawful, who was going to care? There was only one Crown lands ranger with a horse along the entire north coast, and he had his hands full keeping track of the rafts of cedar moving down the rivers, from the Daintree to the Johnstone.

A government bill to raise the red-cedar export duty was labelled a 'lunatic southern proposal',[24] and led to an uproar from the north about it being an affront to free trade. Premier Thomas McIlwraith, concerned for his own seat in Mulgrave, dumped the bill in 1882, leaving forestry working under laissez-faire—'anything goes'—rules. Upwards of 15 million super feet were felled. The Barron River was so clogged that in 1887 the postman couldn't cross it.[25] But the cedar-cutters continued to brand and stockpile their logs in the rainforest in anticipation of a promised railway, and the anticipated gigantic timber traffic.

Armies of cedar-cutters were employed by intercolonial firms to cut millions of super feet, even though there was no cheap method of getting them to market. Teams of bullocks and horses dragged logs at a speed of 8 kilometres a day, each load needing to be rebalanced with any change in grade. Crawling down the escarpment near Port Douglas along the bump track was a major feat. Floating logs down the Barron while it was in flood saw more than a million super feet of red-cedar logs smashed to pieces as they barrelled over the 230-metre Barron Falls. The packers preferred the quicker returns of carting ore, and so thousands of logs were left to rot without a market.

The main pack-supply track that scaled the scarps behind Cairns was almost impossible for timber. It connected Cairns to Herberton via wagon roads that followed the open pockets established by Aboriginal people thousands of years before. Early survey maps consistently contrasted the 'dense tropical jungle' with these peculiar pockets variously recorded as grassed forests of gum and bloodwood, or swampy plains.[26] The ominously named Skull Pocket was reached after ascending the escarpment from the Mulgrave River on the coast, followed by Boar Pocket, Russell Pocket, Allumbah Pocket, Ball Pocket, Pinnacle Pocket, Dingo Pocket and Prior's Pocket.

———

Christie Palmerston was perhaps the first European to suspect that pockets had been created and managed by rainforest peoples. Palmerston's diary of October and November 1880 was written in the precipitous country of the upper Daintree, and describes an intricate landscape with serrated summits, sometimes clothed in jungle, but often enlivened with pockets of bright green.

What set Palmerston apart from the other prospectors was his comprehensive diarising, and his ability to speak one of the languages of the northern rainforest peoples. He was an indefatigable explorer, depending for survival on his Aboriginal companions and carriers (sometimes using handcuffs to control them) and his own marksmanship. Once away from 'civilisation', he shed his boots and trousers for ease of movement. He never wore a hat. He carried a Snider rifle and admitted to finding fulfilment and a measure of fame in treading where no European had trodden before, and enduring extremes, he said, similar to the explorations of the Antarctic or Arabia. He was also known to have murdered over 100 Aboriginal people, in addition to many Chinese people, whom he loathed.[27]

His diaries are jarring. Descriptions of his unsparing exercise of the Snider on people are interspersed with glowing descriptions of the scrub and bush foods. He described beans and quandong and, in one instance, a beautiful fruit shaped like a large apple but plum-coloured, with large seeds surrounded by a soft and pulpy mass that tasted a little like a lemon—the mangosteen.

In the hinterland of the Daintree, the flats were richly grassed and there were expanses of open lightly timbered country. His tasty horses were a lure for groups of people packed into the rainforest.[28] One time, he barely escaped a scrub knife being flung at him;

it landed, shuddering, in the tree directly in front. The knife was made from a pit saw used by cedar-getters. In the nearby camp, he found mauls, wedges and lumps of iron. People encircled him with fire to try to burn him out. But he was persistent.

In 1882, Palmerston was commissioned to search for a railway route over the mountainous rampart west of Port Douglas and Cairns. Nosing along the rampart's rumpled base, following the open pockets, he noted the destruction caused by the immense quantities of red cedar being taken. Near Palm Beach, in May, he saw many people milling about three new canoes and mused that the whites had taken over the land to the north around Port Douglas, pushing people south. With so many people, Palmerston's rifle was used with deadly effect. In July, he was at The Fishery, a place on the Mulgrave River where people had made extensive fish traps. He was attacked, spears flying from both sides of the river. While he 'opened their black hides as fast as he could',[29] he observed matter-of-factly that it was no surprise that they were upset given that a sugar plantation was in preparation, with 900 acres of scrub felled and burned by Chinese workers.

It was a busy year for Palmerston, and in November and December he embarked on what is probably his most famous expedition, blazing a trail from Mourilyan on the coast south of Innisfail, to Herberton. He did it in 12 days one way and ten days in the other without any deprivation, boasting of his achievement when compared with an attempt made earlier in the year by Sub-Inspector Alexander Douglas that took a month and resulted in sickness and near starvation of the party.[30]

By 1882, there were 90 tin claims around Herberton, usually Scottish or Irish miners living in bedraggled huts made from bark that had been stripped from trees along the Wild River. They dug like demented wombats, clinging to single-stick ladders that hung limply against the sides of shafts. As many as 50 drayhorse and packhorse teams loaded with tin left Herberton each day to trundle along the bush-pub clotted road and clamber down the steep escarpment to Port Douglas.[31] A track to Mourilyan would halve the distance.

Palmerston's route lay along a narrow ridge, the route of the Palmerston Highway today, mostly following the well-beaten tracks that people had made. He wrote of the large numbers of people in the forest. He found freshly painted shields drying along the riverbank in the sun, and baskets overflowing with vegetables and nuts, bushels of red berries, newly crushed nut meal piled high on blankets beaten from bark, and fishing nets fixed cleverly on small round hoops or in the shape of a heart. On the edge of the Atherton Tablelands, he camped on Beatrice Creek, a major tributary of the Johnstone River, and walked up to his ankles in crackling nut shells from the groves of nut trees. He crossed a large Aboriginal road approximating the Millaa Millaa–Malanda Road and then went straight as an arrow towards Herberton up what is now Kenny Road, a short-cut route I also use when travelling home from Innisfail. Here, he passed several unscathed red-cedar trees. Travelling up the Malanda volcano, on which our house sits near the peak, he saw large red-cedar and pine logs already cut.

On his return journey, Palmerston came across a group of men quietly gathered on the edge of a thunderous waterfall, each with his own little fire. Across the chasm, the cliffs were streaming with waterfalls, and echoes rolled up from the large rapids of the North Johnstone below. It was a mesmerising sight. It seems that he may have encroached upon preparations for a ceremony. On seeing Palmerston, the men fled into the bordering rainforest, only to appear further upstream with spears and shields, ready to do battle. He drilled bullets through their shields as easily as if they were paper and, with a companion who wielded a long scrub knife, caused havoc before the men gave way. He saw a little boy running away. Catching up with him, Palmerston laid his rifle against the boy's neck and shoved him over. The boy fought, biting and spitting. And no wonder, given the demonic sight of a white man in just a shirt and cartridge belt, his legs splattered with blood. Palmerston

203

called it a day, camping close to the dead men on the brink of the waterfall, today's Tchupala Falls.[32]

———

Until the early 1880s, the only agriculture on the Atherton Table-lands was small Chinese market gardens supplementing the meagre supply of maize and vegetables from Cairns. In 1882, the government threw open 50,000 hectares of the Tablelands for closer settlement, mainly on the edge of the rainforest and along the Port Douglas track, where horse and bullock teams were dragging machinery for the mines as well as corrugated iron in the shape of an A across the backs of mules.[33]

The idea of closer settlement had been forged on fictitious notions of cherished rural landscapes like those in Europe that were ordered, settled and civilised. If the blast of the red-cedar rush shook the forests of the Tablelands, then it was the closer-settlement policy that would bring them crashing down. The bringing of civilisation and the settling of yeomanry in their smiling homesteads,[34] with their rosy-cheeked children and plump wives, meant that the wild, primeval forests had to go.

The problem was that the rough, mostly uneducated European pioneers didn't have the skills to farm. Nor did they want to gain them. Instead, they were chasing quick wealth by logging red cedar or grubbing for gold on the Palmer and Hodgkinson goldfields. Here, they competed with generally hardworking, industrious and frugal Chinese who were, accordingly, deeply distrusted. Racist agitation against Chinese people saw them ultimately excluded from the goldfields. Many found their way to the Atherton Tablelands.

The *Land Act 1886* was drawn up to make it easier for hopeful Europeans of limited capital to take up small selections. Agricultural land could be leased for 50 years. Or it could be converted to freehold at any stage if it was partly cleared and enclosed by a vermin-proof fence of slab timber and occupied continuously by the selector, or his agent, for a five-year residential period.[35] The energetic image of 'throwing open' the land greatly exaggerated the genuine demand. With the majestic rainforest dwarfing bullock drays that looked like tiny toys, the job of clearing was a heroic effort. After burning the felled timber, it was difficult to put even a pick in the ground without striking a network of living roots. Sowing could only be done with a hoe or, ironically, with a digging stick among the roots or the weeds that instantly sprang up. The erection of the required marsupial-proof fence needed about 4000 slabs placed in a trench 15 centimetres deep hacked out of the masses of roots. The task of clearing a whole selection of 65 hectares could amount to a decade of continual manual labour.[36]

In 1886, the decision was made to construct a railway from Cairns to the Atherton Tablelands. Thousands of men toiled at cutting a track up the sheer Barron gorge, boring 15 tunnels, building five bridges and watching a mule a day plummet off the cliffs. For all this, by 1888, a listless 29 leasehold selections and 186 freehold farms had been taken up on the Tablelands.

It was Chinese people, banished from mining, who ran the farms: they were offered five-year leases by European selectors who found Chinese rent quite lucrative, particularly after removal of any valuable timber from the selection. While this contradicted land laws that were designed to exclude the Chinese and to encourage Europeans, no one seemed to bring it to the attention of the government in Brisbane, 1600 kilometres away.[37] The selectors, speculating on the value of the land once the railway arrived, worked instead as cedar-getters, miners, packers or storemen.

There was also another problem: the idea of the bounteous fertility of the rainforest soils was based almost entirely on vegetation cover—and proved inaccurate. Once the rainforest litter was removed, the soils were quickly depleted. Agriculture around Cooktown had already failed by 1886. On the Atherton Tablelands, maize—the poor man's crop—was planted among the stumps,

There was also some small-scale experimentation: sweet and English potatoes, chokos, oranges, lemons, limes, apples, peaches, apricots, bananas, pineapples, grapes, wheat and sugar cane.

But things did not go well. Maize was destroyed by blight, and an army of caterpillars, fruit flies and other insects attacked the orange groves. Crops were planted at the wrong time, and a prolonged wet season in 1886 spoiled the hay. With limited agricultural knowledge, few knew to add lime and sand to prevent the soil from baking hard when it dried.[38] Finally, the rainforest peoples, struggling to find enough food, discovered an appetite for roasted maize as did the possums, wallabies and white-tailed rats (which were cudgelled for hours by lantern light and taken away by horse and cart[39]), and later the European rat. All 110 farms managed by the Chinese, however, flourished.

While 5000 hectares of forest were laboriously razed on the coast, on the Atherton Tablelands there was an appetite for clearing only 650 hectares. And in 1888, rainforest still occupied two-thirds of the district.[40] The government, not letting truth get in the way of a good story, continued to talk up the limitless potential of the Atherton Tablelands for agriculture, which was 'unsurpassed in Australia for the qualities of its soil, its climate, its wealth of natural timbers, and its suitability for close settlement of the white race'.[41] The reality was that less than ten per cent of land selected for agriculture was cultivated. The immediate area around Atherton, which from time immemorial supported the livelihoods of perhaps 250 Aboriginal people on some 165 square kilometres of rich scrub,[42] was, in 1895, a bleak place with clearings entirely overrun by ink weed. Just about every farmer was bankrupt.[43]

By 1897, the area cultivated on the Atherton Tablelands rose to 1200 hectares, which was a tenth of that selected; the rest was held for speculative purposes. Of the 400 or so landholding Europeans, few made any attempt to farm. Young red cedars were noticed regenerating on the fringe of small clearings, but selectors had no appreciation of timber or restoration practices, so any red cedar found was sold to sawmillers, and the rest burned.

———

As the forests were eaten away, the river valleys denuded, and the more open fringing areas cleared, Aboriginal people were forced to harvest the resources left by the invading miners, timber-getters, farmers and pastoralists. The thoroughness with which settlers and the government went about hunting people meant that there was nowhere they could go in safety. White intrusion from the east and west, and crowding from within, squeezed people into the most rugged, thickly forested and marginal parts of the rainforests. There was no alternative but to raid farms.

Bullocks and cattle were killed, houses were looted and burned, camps were robbed. Selectors and loggers worked with tools or axes in one hand and a gun in the other. Troops of Native Police made little apparent difference. It was impossible to move away from the rain-forest tracks on horseback, and few European officers could muster the extraordinary endurance needed to keep up with the Native Police or Aboriginal trackers. A century after the first European fleet arrived—and nearly half a century after pastoral conquest in much of the rest of the country—resistance to European invasion was still strong in the rainforests. As forest strongholds of the people were destroyed, and their means of obtaining a livelihood became more and more reduced, attacks on Europeans increased.

In 1884, a detachment of Native Police killed a group of inoffensive people at Irvinebank. Because the entirely defenceless old man, two women and a child were known to the European community, the officer and his troops were put on trial. The officer was quickly acquitted amid applause from the European spectators, who felt that a conviction was a threat to their freedom to shoot[44] or to spend a weekend 'potting blacks'.[45] The troopers were held in gaol for a year and then released.[46]

The Irvinebank massacre was, nevertheless, a turning point. With the constant theft of crops in the Atherton Tablelands, and the failure of justice at Irvinebank, a deputation comprising some local Europeans approached the government in 1885 with a radical plan: to feed the starving Aboriginal people.[47] In Cairns, where clearing was escalating, about 100 Aboriginal men had already come into town seeking work. But, on the Atherton Tablelands, a further three years of desultory massacres occurred, and the bloodlust of colonial expansion continued: in Bowen, an Aboriginal woman was whipped by a white man with a stockwhip until her breasts were hanging in shredded pieces; and in Cooktown, troopers wearing bloodied clothes captured a young girl, not even 12, and displayed her naked to the town, where she was sold to a citizen.

Against such an appalling backdrop, the Atherton Tablelands Progress Association once again agitated for a white constable and a tracker to be stationed at Atherton to deal with the 'black question without using severe repressive measures'.[48] Pioneer Atherton settler William Bernard Kelly offered very cheap accommodation for the constable and trackers. The settlers had a specific constable in mind, Charles Hansen. He was known for his violent retaliatory measures, in one instance asking for more rifles in 1888 because two rifles were not enough to teach a group of 200 rainforest people a sufficient lesson. He was, however, a fit bushman and able to scour the rainforests with young Aboriginal men who had started working for selectors. These young and vigorous men had been able to eke out a living in the forest and were working for the selectors for rations and tobacco. They couldn't, however, bring enough food back for the old people, for whom they had responsibility, so they also raided the crops.

At Zeigenbein's Pocket, one of the important ceremonial pockets stretching across the Atherton Tablelands, Hansen came across camps with large quantities of maize that had become a staple for hungry people. He broke up several camps, but instead of killing people, he brought them into Atherton with a promise of flour, sugar, tea, beef, tobacco and sweet potatoes at Kelly's store. Some of them took convincing and asked for coastal relatives to come and confirm that the Europeans were not going to kill them. By 1889, people had begun to 'come in' and, despite the government churlishly resisting to pay for rations, the scheme to feed the people and provide blankets was working.

By 1890, with Atherton farmers offering meagre terms of peace that included distribution of rations in exchange for cheap labour, there were 300 rainforest people camping in fringe camps, and about 100 of them were employed by settlers. By 1892, there were 400 people being controlled by rations. To keep the services of Aboriginal families, some settlers gave them pieces of timber and sheets of iron to make a shelter.

In the end, it came down to an accounting approach. One sub-inspector estimated a saving of £2000–3000 per year on detachments of Native Police.[49] A year later, the results for the selectors were dramatic, with a 38 per cent increase in area under crop and a 26 per cent increase in area under cultivation. By 1893, there was a threefold increase in the area under cultivation, and rainforest peoples were allowed free movement throughout the district.

With Aboriginal resistance effectively broken, the government offered 'relief' for Aboriginal people in order to ease 'the remnants of the unfortunate race down the slope they were so rapidly descending'.[50] The doomed race theory was still unquestionably and monotonously accepted. Edward Palmer, a northern squatter and member of parliament, even complained about giving people relief because it would preserve their existence. But the sight of starving families, addiction to opium and alcohol, general hopeless depravity and little girls and women kidnapped and forced into prostitution caused several MPs to grumble that Aboriginal protectors should be appointed. They wanted rationing to continue in northern Queensland, and the sale of opium and alcohol to be prohibited. Archibald Meston, a famous authority on Aboriginal people, was asked to report on the issue.

Meston's 1896 report attacked Native Police and settlers alike. He recommended that reserves be created for Aboriginal people, and a 'Chief Protector of Aborigines' be stationed in northern Queensland, with an assistant to look after 'scattered remnants' in the south. Missions were to be used as food-distribution centres. Aboriginal people living in pacified areas who were not gainfully employed must simply get used to it: 'their land had been taken from them on no other title than the law of the strongest, and they must make the best of any alternative the strongest chooses to offer'.[51]

Meston's criticisms of the Native Police, and the suggestion of increasing the number of white policemen, raised the ire of Police Commissioner William Edward Parry-Okeden, who in his rebuttal and championing of the Native Police provided an uncharacteristic public criticism of the Europeans on the northern Queensland frontier:[52]

> To find even a few such [white] men it would be necessary to recruit from the stations of the far North—that is, from a place and from a class where and among whom at the present time are to be found, masquerading under white and yellow skins, some of the blackest scoundrels alive—wretched who have wrought deeds of appalling wickedness and cruelty, and who think it equal good fun to shoot a nigger at sight or to ravish a gin. So long as such villains escape hanging and live in our country, the blacks must be—and shall be, if I have a free hand and my Native Police—protected.[53]

In 1897, the *Aboriginals Protection and Restriction of the Sale of Opium Act* was passed. Based mostly on Meston's recommendations, the Act assuaged nineteenth-century feeling of bloodguilt and replaced it with twentieth-century paternalism for a doomed race: the murderous savages were now instantly deemed faithful, trustworthy and even affectionate—if somewhat feeble-minded.

The Act was the ultimate instrument of control. Persisting well into the twentieth century, it gave the minister power to move Aboriginal people to reserves and from one reserve to another, against their will if need be—like criminals. A certificate of exemption from provisions of the Act could be granted to any 'half-caste', defined as 'any person being the offspring of an aboriginal mother and other than an aboriginal father'.[54]

'Half-castes' worked under an agreement that conveniently did not stipulate remuneration. The effect of the Act was to give the government and employers control of Aboriginal labour. People lost their freedom. Even their economic and social activities were prescribed solely in order not to offend settler interests: Aboriginal people who were deemed offensive to whites were removed, while those who were useful were allowed to stay. Any Aboriginal person living close to white settlement had to scramble for a role that was deemed suitable for them by white society, from whom there was an expectation of complete submission. The alternative was to go to a reserve. On reserves, people could be imprisoned for any crime or misdemeanour. No ceremonies or traditional customs were permitted. There were no civil rights for Aboriginal people, who were declared 'inmates in perpetuity'.[55]

Thiaki's people were driven to the precipice of their existence and to the very edge of physical and psychological destruction.[56] How Thiaki mourns the profound and instantaneous lack of coevolutionary communication, I can't begin to imagine. The loss of careful management, maintenance and creation of niches that supercharged life must have left an impact I don't know how to define, let alone measure. But I see it while walking on an old Aboriginal footpath through the forest in the often-impenetrable layer of undergrowth, the lantana-choked streams, the vine-choked forest edges, and the loss of species such as the spotted-tail quoll and cassowary. The biggest loss is of the giant trees, which once dominated canopies and choreographed the structure, dynamism and diversity of the rainforest. When they are taken out, holes are ripped in the canopy exposing the forest to direct and scattered light, wind and weeds.[57] Thiaki and their people had no choice but to remake a different world from the fugitive pieces.

# CHAPTER THIRTEEN

# Fugitive pieces

What is a tree? Is a tree just a lump of wood crowned by leaves? A tree is truth with a nail in its side, beauty built out of air and stones—mellow in death, lively in spite of human foolishness. Trees adore wind and birds, sunlight and sweet air, kookaburras and cows, raindrops in quiet evenings. Nobody but a plant can dine so noiselessly, and on such strange foods as the invisible carbon of the air and the harsh minerals of the ground. Nobody but a tree can grow so huge—not even whales and dinosaurs. Nobody but a plant has green blood to capture the energy of the sun. Trees have no blood banks to succour them after fire and mutilation, yet without the green stuff of their sap, there would be no redness in animal blood, no sun's energy and no life for us who cannot dine on dust. A tree is a magic creature, whose ancestors are lost in the mists of time . . .

Len Webb, 'Trees are your friends'[1]

The forest, no longer properly cared for, started sickening. Large areas were opened up to strong winds and were damaged by cyclones. Scrubby vegetation, low bushes and tangled undergrowth invaded what had been a landscape of tall trees and dappled earth with a sprinkle of understorey. Organised shooting parties exterminated scrub turkeys that had been the pest controllers of borers and other insects in local areas. Any alienated land continued to be taken by speculators, who reaped profits from the saleable trees.

The railway reached Mareeba in 1893. By 1897, over a million super feet of red cedar and kauri had been dragged through the scrub and loaded onto trains[2] for cigar boxes in the London market. In 1900, the first forest ranger, F.W. Lade, was sent to the Barron Valley for three months to conduct a survey of timber resources. He found decaying red-cedar logs lying all over the forest: the market for timber was clearly not as plentiful as imagined. Lade found that the so-called inexhaustible resources of less than 20 years before were almost cut out, and that 20 million super feet were left to rot in the Upper Barron near Thiaki. In just one place, he found 400 red-cedar logs lying cut but worthless.

Caught up in the economic global depression of the 1890s, it was clear the government had no idea what had been going on in north Queensland. After Lade reported his findings to the government, cedar-cutting licences were abolished. Instead, a log-based royalty system was introduced to slow the rate of felling. It didn't work: the royalty was trifling and could be avoided.

Atherton was a rough and ready outpost of four hotels and six stores. The one main crop, maize—which fuelled the horse and bullock teams—was controlled by the Chinese. Distance to markets meant that agriculture export was all but doomed. Even vegetables needed to be imported from Victoria. Speculators, however, continued to talk up the value of the land. Propaganda pamphlets issued to new immigrants arriving in Queensland in 1905 proclaimed that there was 'room for four million white Australians, and a paltry few thousand Chinamen need not frighten them'.[3] One visitor from Lismore, a successful dairy region in New South Wales, found that the largest dairy had 26 cows.[4] Of the first dairy herds driven from northern New South Wales to the Atherton Tablelands, a distance of 3000 kilometres, only a quarter survived the trip. Isolation, global economic depression and finally the cattle tick all but wiped out the early dairy industry.[5]

A brief tin and timber boom in the early 1900s, together with healthy supplies of Chinese-grown maize (there was not a yeoman farmer to be found), resulted in a further length of rail being laid to Atherton from Mareeba in 1903.[6] Governor Sir Herbert Chermside opened the railway with fanfare, comparing the forest clearing at Atherton with the forest clearings of the Caucasus, which, he said, was the birthplace of the white races. Atherton, he claimed, had 'all the factors which make for the prosperous settlement of the white races'.[7] Even though that prosperity was created by rainforest peoples and the Chinese population.

Timber was earning double that of ore and maize, and it carried the Atherton Tablelands economy:[8] by weight, it was 90 per cent of goods transported by rail in 1904. Red cedar alone accounted for 80 per cent of Queensland's timber exports, more than half coming from Atherton. An accurate assessment of the value of the timber is impossible because no records were kept. The largest logs were reserved for export in the belief that they would gain the highest price. But they were so huge that they needed to be split by dynamite to pass through railway tunnels. In London, they were too big for the English sawmills. This didn't stop cedar-cutters from dropping the biggest trees.

The forest was giving up its final secrets: 2 kilometres from Thiaki, Hypipamee—a spectacular volcanic diatreme—was newly discovered by Europeans, and made it onto the 1904 topographic map.

———

215

The idea of illimitable forests was taking a beating. Though 70 species were suitable for milling, only about five were used to any extent: *Castanospermum australe* (black bean), *Gmelina fasciculiflora* (northern white beech), *Agathis robusta* (kauri pine), *Cardwellia sublimis* (northern silky oak) and *Flindersia brayleyana* (Queensland maple). In 1892, colonial botanist Frederick Manson Bailey complained of the short-sightedness of throwing money at mineral wealth without a thought for the riches of the trees that, he said, were of much more importance to human wellbeing.[9] Scorn met this crazy and esoteric idea. For his efforts, the position of colonial botanist was abolished. Bailey carried on work for no salary. In 1904, he complained that the Australian vegetation might follow the Aboriginal people and disappear.

Sawmillers, who had introduced the revolutionary technology of steam milling and kiln drying, also began to see the supply of timber being threatened by clearing for closer settlement and agriculture. They wanted the forest conserved for the use of the sawmillers and expected the state to be responsible for forest restoration.

Pressure to better manage Queensland's forests came from unexpected directions. The earliest call for conservation was, paradoxically, made by the botanist on Dalrymple's 1873 expedition. Walter Hill claimed that 'a half million acres' of good land around the Johnstone River was the most valuable discovery in Australia for cultivation.[10] Two years later, in 1875, he poignantly catalogued the forest abuses of the colony, drawing particular attention to north Queensland with a plea for immediate steps to conserve the timber along what is now the Wet Tropics coastline.[11]

Hill's efforts in 1875 led to the formation of a select committee to investigate the matter led by John Douglas MLA. In the subsequent report, Hill recommended reservation of state forests for reafforestation and for scientific research. A timber export duty was to pay for these. The government was mute on the idea. You might have thought that the elevation of John Douglas to premier in 1877 would have had a positive effect. It didn't. He was quicky ousted by conservative capitalist and bully[12] Thomas McIlwraith in 1879.

Elsewhere, new ideas on environmental management were emerging, including flora and fauna protection, agitation for wiser use of forests, and concepts of sustainable yield—though how they would be applied in Australia was anyone's guess. In 1879, wealthy pastoralist Robert Martin Collins, who had returned from a visit to Yellowstone National Park (the world's first national park), lobbied for the protection of the wild and beautiful McPherson Range on the border between Queensland and New South Wales. In 1889, a sawmiller and member of parliament, Richard Mathews Hyne, was concerned about the encroachment of agriculture on the timber industry. He proposed permanent forest reserves and the creation of a forestry department.[13] It was the same year that Archibald McDowall, the surveyor-general, made a prescient statement:

> It can hardly be questioned that the time is approaching when the wholesale destruction of timber in many parts of the Colony—much of it of a wantonly wasted nature— will be severely felt. Suddenly, when the depredations of a careless population have produced the inevitable result, the subject of forest conservancy will assume a prominence not yet accorded to it, and it will be a matter of general wonder that our short-sightedness did not allow us to realise that destruction without replenishment must lead to scarcity.[14]

It was to no avail: the funds set aside by the government for forestry in Queensland in 1891 amounted to £65 (about $16,000 in today's value).

In 1897, a commission on land settlement gathered extensive evidence of continued waste of forests. It also recommended the establishment of a department of forestry along scientific lines to promote afforestation. But in 1900, the commission ushered in not an independent department of forestry but merely a forestry branch within the overarching Lands Department that oversaw closer settlement. The branch, which was to service the entire state, had a staff of three: a director of forests, and two forest rangers (including F.W. Lade). Its job, which had nothing to do with conserving forests, was mainly to collect the small revenue from the logging licences; to determine how much timber could be logged in the 660,000 hectares of forestry reserves; and to recommend other forestry reserves. Forest reservation was to be strictly residual to farming. There was no sign of any forestry legislation.

Agitation from a small but articulate minority continued, however, and this finally ushered in the *Queensland State Forests and National Parks Act* in 1907: uniquely in Australia, national parks came under the control of the forestry branch.[15] Forestry, however, was still subservient to the Lands Department. But the Act proved to be a hollow shell because its purpose of guaranteeing future timber supplies through forest management and protection was left to regulations. These would take Norman Jolly, the director of forests, seven years to draft; in the meantime, there would be a policy vacuum.

Into this policy black hole, like paper into a shredder, were fed the Atherton Tablelands forests. Free of any legislative constraints, every selector became, by necessity, an amateur timber dealer. And many of these were swindled by professional ones. By 1914, it was hard to find a standing red-cedar tree. The forest—including its valuable timber and its incomparable biodiversity—went down hectare by hectare and was torched. It was a forest cluster fuck of the first magnitude.

There was no perception of the importance of natural resources, no appreciation of conservation principles, no idea whatsoever of wise use. Jolly's successor, Edward Swain, was to say that there was 'no policy at all except to parcel out the country into individual blocks regardless of the consequences or of topography, and to tab each with an owner's name, regardless of his . . . bona fides'.[16] The Act made no effort to resolve the growing conflict between closer settlement and forestry. Indeed, most of the efforts of the two rangers was in excising areas from existing timber reserves that were suitable for closer settlement.

Even though it was already clear that closer settlement was not working, amendments to the *Land Act 1910* cemented closer settlement as priority policy. This would remain unchanged for 50 years, no matter which political party was in power: the difference between political parties was who would open the most land. In 1910, a correspondent for the *Cairns Morning Post* was concerned that rainfall would be diminished as the land was denuded of rainforest, and the day would come when trees would need to be planted.[17]

But the pressure to open country was relentless. Surveyors on the Atherton Tablelands surveyed up the precipitous sides of mountains such as Mount Father Clancy. The steep cone of volcanic Mount Quincan, a prominent feature seen from Thiaki—and an important sacred site—was divided into three dairy farms. Only on the precipitous escarpment at Gadgarra and the Lamb Range—where farming was impossible—were forestry reserves proposed.[18] A small Crown lands reserve around Lake Barrine was proclaimed, but this came too late since most of the big trees had been removed.

The Atherton Tablelands gathered a reputation for champion axe-wielders. As the forest went up in smoke, the fires were sufficiently awe-inspiring to attract local criticisms of the valuable timber being hacked down and burned for a literally worthless crop of maize.

———

As the twentieth century progressed, three main groups, not mutually exclusive, took up land on the Atherton Tablelands: the experienced farmers, such as those who first took up land around today's Malanda; the speculators; and the timber interests. Between 1907 and 1914, more liberal land laws were enacted to encourage the first group. The *Group Settlement Act 1907*, for instance, allowed groups of settlers to join and register for group application to buy land. In 1907, 13,000 hectares were open for group selection. The land was sold at today's equivalent price of $1140 per hectare.[19] This included the timber, which could be sold by the group. Setting up a dairy (a cost of £214, or around $50,000 in today's money) or clearing the land (£4 per acre,[20] or around $2000 per hectare in today's money), required settlers to have capital behind them. Many didn't, and they sold out after taking the timber. Or forfeited the land by walking off.

One of the first groups of settlers, James English and his family, came to Malanda in 1907 from the Richmond River district of the Big Scrub in northern New South Wales, having migrated there from the smaller scrub on the south coast of the state. They came to the Atherton Tablelands with a third episode of scrub-conquering in mind. The English family selected their blocks sight unseen from the maps of the first surveyor of the area, Herbert Maguire (who also surveyed Thiaki), which were circulating the dairy regions of New South Wales. The English family chose blocks described as 'river flat, heavily timbered, red beech, vine scrub etc.'[21] to establish their dairy. They cleared 12 hectares of rainforest in 1908 and sowed it with paspalum for grazing.

The first 12 hectares of forest around Malanda had been felled just months earlier by Jack Prince and his offsider, Billy Barlow, an Aboriginal rainforest man whose murdered family probably once wandered under the closed cover of Thiaki. Billy was taken, according to local rumours, from his dead mother's breast in the mid-1880s after a punitive police raid, and raised by one of the first settlers' wives, who suckled him alongside her own newborn child.[22] Billy learned expert skills from being raised into a sawmilling family. He became a famous axeman and well-known champion woodchopper. He was also one of the first to haul logs with a steam tractor; some of the logs went into building the Malanda Pub, owned by the English family. The first settlers needed Aboriginal labour to fell trees on their blocks, and many young Aboriginal men chopped down their ancestral home while helping settlers clear the land.

———

One day I took a copy of the original *Deed of Grant of Land* for two of Thiaki's lots to the ever-helpful local Eacham Historical Society based in Malanda. They went through the list of Thiaki's selectors with me, pulling out a weighty volume of land titles filled with names inscribed in perfect cursive. Thiaki passed hands almost yearly from 1910 to 1916 as first one logger then another took their pick of trees, and then transferred the forest on to the next ardent logger. After 1924, one of the selectors cleared and grassed about 30 hectares along the main valley. This improvement allowed Thiaki to become freehold in 1930.

The first aerial photograph of Thiaki was taken in 1943. It shows that the 30 hectares was cleared right down to the banks of what would have been a clear, gurgling, first-order stream at the base of 40-degree slopes. I'm told by local foresters that highly flammable *Melinis minutiflora* (molasses grass) was planted to carry fire right down to the water's edge. Today, this grass remains in scabby patches, like unhealed wounds, on Thiaki's steepest slopes.

———

I struggle to understand what was gained by any of this maniacal destruction. The drive to closely settle the north in the name of White Australia xenophobia not only caused the destruction of rainforest and the rainforest peoples, but also facilitated the pauperisation of the dairy industry. For years, farmers paid less for butter and cheese than it cost to produce. As war loomed in Europe, the government imposed low prices on butter, a key export item and main dairy product of the Atherton Tablelands, to help war-torn Britain. Competition from other countries, and brutal butter cartels, meant that most dairy farmers couldn't make a living. They converted to maize, sold up or simply walked off.

By 1912, only Chinese farmers—with their farming knowledge, industry and organisation—had demonstrated that maize was viable. Industrious Chinese made inept Europeans uncomfortable. Racist settlers questioned the suitability of Atherton for closer settlement of the white races—the desired path under the White Australia policy—while Chinese people were allowed to prosper. It was not a sentiment held by all settlers, particularly those who were receiving Chinese rent, but it was a sentiment that won out. Chinese people were debarred by law from agriculture. Since they were also debarred from mining, labouring and just about everything else, they had little alternative but to leave. To this day, there are no Chinese-run farms on the Atherton Tablelands.

The leases were instead resumed in 1919 for the post-World War I soldier settlement scheme. This disastrous policy replaced competent agriculturalists with soldier settlers who were often city-reared, suffered from war wounds or post-traumatic stress (or both), and had no farming experience. To add insult to injury, the soldier settlers were advanced large loans that they couldn't pay off. Virtually set up to fail, most farmers were reduced to grinding poverty. Some committed suicide, and others walked off with their families, broken, poor and hungry. Much of the nearly 6000 hectares set aside for returned soldiers was invaded with weeds and washed of topsoil. The shire council was unable to collect rates, and roads deteriorated.

Nevertheless, 1920s post-war revitalisation brought with it new visions of national development. And in the north that meant filling all of the 'empty' spaces. Australia's future population was exuberantly estimated to reach hundreds of millions.[23] On the Atherton Tablelands, the vision was a mirage: of the 100,000 hectares selected, only six per cent were cultivated and the rest was abandoned to inkweed, milkweed, wild tobacco and stinking roger. The Cairns land commissioner dismissively waved this aside, saying that the area simply didn't have the right class of settler. The steep slopes and poor soils of nearby Mount Molloy, the headwater catchment of the Mitchell River that flows into the Gulf of Carpentaria, were opened for the army of stout settlers who would supposedly flock there. Inevitably, dairying proved a failure, and the Mount Molloy rainforest was replaced with degraded grassland.[24]

North Queensland's first sociological study,[25] done in 1925, revealed that those with the worst living standards were dairy farmers of the Atherton Tablelands. Most houses had no water or bathing facilities, and there were signs of neglect in children.

Undeterred, the ludicrous baying for more land to be opened continued even though there were no suitable blocks or markets. Not surprisingly, the clamour did not come from buyers who wanted to live on and work the land, but from speculators who wanted more 'woodenheads'[26] (stupid people) to live on it. Broken and unworkable blocks of land were sold as incomparable dairy paradises.

Under such intense local pressure, the rough rainforest-clad land between the Russell and North Johnstone Rivers at Boonjie was opened in 1924. The official estimates of timber were held low, and the rent was kept cheap. Timber traders scrambled to the blocks: 22 of the 42 blocks were rushed, and the blocks were subsequently forfeited. When later settlers came, they discovered that they were paying rent on timber that was no longer there. Nothing had been learned from the past openings. And most of the blocks were left for lantana.

By the 1930s, Atherton Tablelands pastures, stripped of rainforest, were in a deplorable state. A native scarab beetle, *Lepidiota caudata*, found the new pastures a haven for its larval stage, and over 10,000 hectares of pastures were ravaged while weakened cattle succumbed to pleuropneumonia. Like breaking a bone to take your mind off a sprained ankle, 20,000 cane toads were released that ate everything except the beetle.[27]

The ideology of the yeoman farmer—sturdy and independent, with a clutch of rosy-cheeked children plus a plump and happy wife—never eventuated. Poverty, ignorance, misery and land degradation were the outcomes. It is true, however, that northern Australia was never held hostage by the 'yellow hordes'.[28]

———

The only voice to question the destructive mania of land clearing in northern Australia was Edward Swain, who had replaced Norman Jolly as director of forests in 1918. In 1930, he stated bluntly that foresters had spent valuable time and money on wild goose chases in search of thousands of acres of fertile soil suitable for closer settlement and allegedly locked up in state forests only to find them existing in the imagination of progress associations.[29]

Like Jolly (who was a Rhodes Scholar), Swain was a professional forester. A legend in forestry to this day, he was outspoken and uncompromising. He did not suffer fools, and the dominant view of closer settlement was foolish. Swain pointed out that forestry was also land settlement and, unlike attempts at agriculture in the region, it could support a local economy in perpetuity. Swain lobbied for an independent Forestry Department, and in 1924 he finally achieved the separation of Forestry from the Lands Department. But instead of being a truly independent department, the government formed the three-man Provisional Forestry Board (with Swain as chairman), which answered directly to the Minister for Lands.

Nevertheless, under that structure Swain was able to expand Jolly's small team of trained professionals to build an industry streamlined into sylviculture, forest management and economics. This approach, which is still in use today, soon provided the state government with good revenues. But the government, with opposition from sawmillers and politicians alike, was reluctant to reinvest any of these revenues into forest management. The argument was baffling: that logging restrictions and royalties were inimical to the public good service of forestry. This public good seemed to be defined as the provision, on demand, of a limitless supply of timber.[30]

Even more baffling, forest management was seen to be against the freedom of timber-getters to operate as they saw fit—outside the law. The Forestry Department, so it went, locked up the land and hampered settlement.

Queensland anti-intellectualism was also on steroids. Lands Department bureaucrats, who had no training in any field, treated professionally trained foresters with derision. Science, bureaucrats said, got in the way of common-sense management. The practice of sending forestry officers to the United States for training was also nonsense because 'American training unfits an officer for practical forestry administration in Queensland'.[31] The conflict came to a head with the formation of the 1931 Royal Commission on the Development of North Queensland, which was tasked with assessing the two options: closer settlement and forestry.

It soon became clear, however, that the royal commission was no more than an inquisition against forestry and Swain in particular.[32] The three commissioners, chest-thumping close-settlement advocates and miners made sure that the Forestry Department—which had no representation—got short shrift. Hearsay evidence of timber-getters was dealt with as if it were that of qualified foresters. Under these circumstances, it was easy to dismiss evidence from the Forestry Department of frequent timber trafficking. The testimony of timber-getter A.E. Turner, who had an uncannily detailed knowledge of timber taken from Boonjie properties, became the basis of a new policy to open even more blocks on Boonjie. Turner enthused that Boonjie would be the best dairy land in the Atherton Tablelands. Today, stripped of its trees, this rugged and steep land is among the most degraded in the region.

The submission of another timber merchant to the royal commission was crowned with its own appendix, which concluded that if a holocaust were to raze 90 per cent of the forest on the Atherton Tablelands (as it indeed has), 'it would be a far greater godsend to the selectors, to North Queensland and, vicariously, to the State, than is ever at all likely or even possible to be achieved by the most "enlightened forest policy" imaginable'.[33] Jaw-droppingly, the commissioners argued that there was an abundance of timber—in fact, too many trees. The commissioners even denied that erosion had occurred following clearing, saying it might be correct to a limited extent, 'but not to a degree that need cause any concern'.[34] They were particularly critical of the Forestry Department for not allowing the removal of dead or dying trees and for harbouring weeds and vermin.

Fired up in zealous righteousness, the commissioners emphatically pronounced that north Queensland was the testing grounds of the European race. It was where 'White Australia must stand or fall'.[35] The commissioners recommended continued forest clearing—demanding that the dairy industry be expanded—and a general opening of all lands except inaccessible mountains and gorges. To not do that, they said, would mean the end of closer settlement 'and eventually the overthrow of our social order'.[36] Forest management was regarded 'as alien to the British instinct of colonisation, and to every sound principle of progress'.[37] Furthermore, the Forestry Department had strangled land settlement and threatened the stagnation of the north because of its unpersuasive idea of providing timber supplies for future generations. With evangelistic conviction, the commissioners held that the Forestry Department should provide as much timber as the public wanted. This public good was its main job. The idea that forest management could be made a profitable enterprise and compete with other industries was considered simply unsound.

The Forestry Department was subsequently reabsorbed under one Land authority, which would reassert the moral superiority of closer settlement. The forests, it seemed, had to go. While Swain was instrumental in stalling the deforestation of residual prime forest in north Queensland, it came at a personal cost. He was removed from office in 1932.

———

Forest reservation was not preservation. For all of Swain's passion, his view of forest management was for the attainment of a perfect forest where the multiplicity of species was arranged and ordered into millable trees producing high-grade wood. This meant that non-commercial 'weed' species were brushed, ringbarked or poisoned, and what was left was to be spaced according to a formula. Such a tree farm was an impossibility in the Wet Tropics, with its teeming population of trees. But by 1960, nearly 1000 hectares had under-gone this sort of treatment in north Queensland. The forest, it seems, had other ideas. Spontaneous regeneration quickly swamped the effects of these forest-management practices, which were abandoned in the 1970s.[38]

The idea of retaining any forests as national parks was also not on Swain's agenda.[39] Indeed, it would have been perceived as ludi-crous in the 1920s to argue for forests to be left in the unproductive state. Swain's important legacy, however, was to stall the clearing of forests.

In the world, the concept of sustained yield—only cutting as much timber as the forest grows—was by now a well-established concept. It was ignored in Queensland. Between Swain's 1932 sacking and 1960, the number of trees felled quadrupled, especially after the introduction of mechanical logging. There was one attempt to bring this under control in 1936 by introducing legislation that required sawmills to be licensed. But this backfired when licences were issued to anyone who applied for one: 600 sawmills clamoured for timber from forests that could not support such an insatiable demand.

The onset of World War II meant that forests were hacked even more mercilessly. And post-war reconstruction saw the govern-ment ordering the Forestry Department to respond exclusively to community demand, which itself was whetted by log prices set below market values. In 1948, a post-war log quota for sawmills was pegged on the maximum quantity of logs that sawmills could take. Perversely, this quota came to be accepted as the sustained yield. Victor Grenning, who replaced Swain as director of forests, reminded policymakers that trees didn't grow overnight. He also began voicing the rising awareness of forests as places of beauty, pointing out the strange and handsome plants and animals that would become extinct without the forests.[40] Thanks to the Forestry Department, national parks had increased in number, but these were all small areas around scenic spots, or rugged areas of scenic attraction that had already been logged. They were not places for conservation of natural areas.

In 1952, when the forests crumbled under the demands of nearly 1300 sawmills, the laissez-faire licensing policy was finally discontinued. Several years later, the Forestry Department warned that the licensed capacity was still more than twice the annual cut.[41] In 1959, the Forestry Department finally got its own independent legislation to control the permanent reservation of forestry areas. While there was no voice for other values, across the country sentiment was building for forests to be accepted as an essential part of the landscape. An exception was the Wet Tropics, where forests were still seen as an obstruction to dairy farms.

———

By 1940, the Atherton Tablelands landscape was one of depleted soils, weed infestation and erosion on steep slopes so severe that fences disappeared under avalanching soil. It resembled 'real Ma and Pa Kettle country', and an air of disaster had descended over it.[42] Landholders pleaded for advice and assistance, resulting in a regional experimental research station being established to investigate pasture and farm management. Eventually, an extension service was created to provide advice at individual whole-farm level on pasture, weed and herd management. Together with government support, this finally began to improve the dairy industry. But it didn't last.

Post-war restructuring shifted focus to efficiency and lower-cost production. Out went the ideology of closer settlement. In a breathtaking reversal, it was replaced by policies of rural reconstruction and farm amalgamation: it was now get big or get out. Or stay impoverished. Once again, the Atherton Tablelands' highly subsidised dairy industry was hit hard. Without government support, farmers walked away from the industry in droves.

You would think that the last thing anyone wanted would be to open and clear more land. But, unbelievably, in 1933—on the eve of a state election that would reap political gain by reaching back to the toxic advice of the 1931 royal commission—24 portions comprising about 2000 hectares were opened for clearance and settlement just south of Thiaki, at Maalan. In a concession to the Forestry Department, timber was to be removed first. This at least slowed down the opening process until World War II brought it to a temporary halt. Even a subsequent soldier settlement program was rejected because of the steepness of the blocks and the threat of soil erosion in a part of the Atherton Tablelands notorious for high rainfall and persistent cloudy and wet conditions: settlers spoke of living among 'square clouds' because they travelled through one window and out the other.[43]

An appraisal of the area by the government through a short-lived post-war Bureau of Investigation[44] noted that it had limited prospects for dairy. The Forestry Department, on the other hand, regarded it as one of the most prolific timber-producing areas left in Queensland. But with an election in the air, the Labor government sought victory by anachronistically talking up pioneering opportunities, even though the concept had pretty much run its course.

Selection of the Maalan lots in 1954 was by ballot. Many selectors were to view success with dismay when they found that their blocks clawed up precipitous slopes that were generally 'up to shit'.[45] For the first two years of the Maalan leases, selectors were not permitted to sell valuable timber. This allowed time for the Forestry Department to remove it first. Any timber remaining could be vested in the selector who, in the meantime, was to clear 40 hectares in the first three years. To produce a good burn required a dry window, usually around December. Advice from old loggers was to smash up the trees as much as possible to get a hot fire.[46] Some settlers felt that it would have been better to reforest the land back to loggable scrub and earn money that way, but this was seen as a ludicrous suggestion. A subsequent productivity survey of the Atherton Tablelands described the Maalan area as degraded,[47] a situation only partially reversed on the flatter blocks with the pasture revolution of the late 1960s.

Agricultural expansion had reached—and, on the Atherton Tablelands, clearly overshot—its limits. In the 1970s, only those efficient and adaptable farmers with capital survived. Others sold out as soon as land prices increased. The devastated Wet Tropics rainforest, ravaged and stripped of its voluminous skirts, hunkered into the refuges of the rugged mountains. There were 1000 dairy farms on the Atherton Tablelands in the 1960s. Widespread amalgamation, attention to better breeding, modern dairy design and the inability to make a living on small, semi-subsistence, family farms without subsidies reduced this number drastically in the 1980s. Deregulation in 2000 reduced them even more; in 2020, there were only 41 dairy suppliers.[48]

In 1966, legendary ecologist and botanist Len Webb put forward a series of proposals to protect the remaining habitats of the Wet Tropics. This was the first reference in the scientific literature to the international significance of the lowland rainforests.

———

The rich yet narrow strip of coastal lowlands on the other side of the rugged behemoth of the Bellenden Ker Range from Thiaki—only 40 kilometres from Thiaki as the crow flies—had been all but cleared. In the 1960s, one large swathe of 13,400 hectares of stunning and irreplaceable lowlands—intricately patterned with tea-tree and paperbark communities, eucalypt forests, palm forests and rainforest—remained on Crown lands. It was the last large area of lowland forest anywhere in the Wet Tropics.

Cattle baron Bob Kleberg of King Ranch in the United States was the world's biggest pastoral landlord in the mid-twentieth century.[49] In 1952, fuelled by easy wealth from oil found on his Texas land, he came to Australia and purchased a tract of land the size of a European country that stretched across the northern savannas.[50] Nobody but the conquering robber baron Kleberg thought that the steaming tropical rainforests were capable of beef fattening. He estimated that he needed around 20,000 hectares to make such a venture workable. In 1963, the Lands Department—still prepared to clear all land where topography permitted—offered the intact lowlands for $5 per acre on the condition that they be cleared, and topped up the offer with already cleared land.

King Ranch used newly available earth-moving machinery to smash through the forests and level the land. Newspapers of the time gushed about the herculean effort of clearing, the dynamiting of every boulder and the filling of every swamp. It would take five years to scrape the land clean of rainforest.

In a rare capture during 1963, linguist Bob Dixon documented a conversation between the rainforest people who were watching King Ranch annihilate their Country.[51] Snatches of the conversation called the names of places as they went under the bulldozer: Dagu (a swampy creek); Banjabanja (a bend in Davidson's Creek); Gugar (a swampy lake) is finished, all the people gone; Wanydyan (another swamp); Giyarra is finished; Gurrungurru (a clearing in the flat land) now has a road lying across it; Dagarrangami (a camping place) is finished; Guwanya (a forest camping place) is being cleared. Look out! They're pushing Yunigali down (a stone that is a dog—children are not allowed to go near it, or else a cyclone will come). Jagurru, the place where lots of umbrella palms grow, is finished; Wulgu (an ancestral place) is cleared; Balara (a camping place alongside the creek) is gone; Juwijan (a swamp) is finished. Gundulu is finished—all the white women and their children are camping there now along the river.

One rainforest family was still camping at Garrngurru. People who wanted to visit were frightened because there was a fence with no gate. What about Yudi-birrbany? It's finished, you're too late. Guynyibila swamp is finished; Galan (a creek) is finished. Your swamp is finished—there's a plain lying there now. Grandfather's place? Yes, it's finished.

King Ranch men told the people to come over for scrub-hen hunting, to follow the bulldozer. But why do that when there was nothing there? Snatches of conversation continued: you have a look, there's nothing. What would you go for? It's strange after it's been cleared, empty and silent. We used to muck around in Bawul swamp. At Jilgarriny waterhole (now a dip yard) we put *gilbay* (foaming bark seed) into the water and picked out the dead fish. Now I've got nothing, just sugar and no money for tobacco. My grandfather is buried over there, but now it's pulling up ferns. Grandfather will have to move further upstream because King Ranch has flattened his burial place.

Today, all that remains of this priceless cultural and biological wealth is the 1700-hectare internationally significant Eubenangee Swamp.

# CHAPTER FOURTEEN

# The final stand?

In the early 1970s, a commune of dropouts, drifters and bohemians washed up at Cedar Bay after being hounded out of Kuranda, the hippie capital of Australia, by police. Cedar Bay is remote and stunningly beautiful. As the crow flies, it is across the Atherton Tablelands from Thiaki, over the tossed landscape of dry rainforest and savanna of the Black Mountain corridor, over the escarpment west of Port Douglas, across the rugged Daintree Rainforest, past Cape Tribulation and the Bloomfield River: 180 kilometres in a straight line. Access is by foot from Bloomfield, or by boat. Clearly, the commune wanted to be left alone, in peace and quiet. But in August 1976, it was raided by police in a military-style operation involving aircraft and naval boats. Twelve partially naked and totally bemused people were arrested, and their fruit trees and homes destroyed. It seems that the free-living vegetarian lifestyle had been an ongoing affront to the conservative cow-eating Queenslanders.

The raid attracted national media attention. When no stash of drugs was found, the charges against the Cedar Bay residents were dropped. But the ham-fisted scandal brought down the Queensland police commissioner and drew the international spotlight to the Wet Tropics rainforests of Australia.[1]

The times they were a-changin'. But not that much. At Cape Tribulation, 1976 brought bulldozers from real estate developer George Quaid to push a road through the last tract of barely explored rainforest between Cape Tribulation and the Bloomfield River. It was a place where the iconic rainforests flowed down from steep escarpments to greet the Great Barrier Reef in over-hanging caresses.

The year 1976 was also when Barry Pember purchased Thiaki. He had made some money on the coast using newly minted heavy machinery to load sugar cane. When he bought Thiaki, it would have been possible to clear 40 hectares in a day using two large bulldozers working in tandem. Thiaki could have been deleted in less than a month.

But Barry bought it for the last of the big trees. Following what seemed to be an established pattern, rainforest man George Davies worked for Barry for the first 15 years, walking the sheer valleys and looking for big trees that might have been left by the parade of loggers in earlier decades. Big trees like my friend, Tree, in the hidden valley, are now rare on Thiaki. When they found a suitable tree, Barry and George would cut a V shape out of one side of the trunk before jumping up on springboards on the other side to chainsaw across the trunk, carefully approaching the V from a few centimetres higher. As the saw cut slowly opened, the tree would moan, and the men would jump down to the ground just in time before the tree began to drop, slowly at first and then picking up speed before hitting the ground with an earth-thumping crash. A shocked silence would be interrupted by a gentle and sorrowful rustling as foliage drifted down from the hole newly ripped into the canopy. The top branches of the felled tree were then sawn away; the decapitated log was chained and clamped, and then dragged by a powerful tractor to Barry's private mill.

Al, one of Barry's three sons, said this livelihood ensured that they never wanted for anything. Each boy, for instance, got a motor-bike. And a motorbike rally called the 'hill climb' was a regular event at Thiaki. One of the steepest slopes of Thiaki was selected to test the skills of the district's boys as they rode their thrash-ing motorbikes up it, out of control, to see who could get to the top in one piece. Barry once rather proudly showed me the large erosion slump in the slope where the hill climb took place. I got to know that slump well, as I scrambled across its uneven and ankle-breaking surface a dozen times, first surveying the line for the plantings, then spraying the grass and weeds, and finally carrying the trays of plants to the planters. Today, it is now permanently anchored by restoration trees.

Al, eyeing another of our revegetated slopes from the verandah, shortly after Barry had passed away, shook his head as he recalled that when he was barely a teenager and this slope was newly cleared for cows, he would, under instruction from his father, take a chainsaw to keep the forest and the weeds at bay. Old aerial photographs show Barry's and George's handiwork as they began to remove more of the forest from the steep slopes, crashing the trees right into the trickling water of the little creek below the platypus dam. Each weekend, Al was sent out to keep the new pastures free of the weeds that were constantly encroaching in a guerrilla war with the grass.

On the Atherton Tablelands, the rainforest now mostly comprises flimsy fragments. Totalling 10,000 hectares, they are found in ragged patches on private or leasehold land on steep or precipitous slopes. Between 1978 and 1988, 1000 hectares[2] of this erosion-prone land—totally useless for agriculture—was cleared, including the 50 hectares or so of Thiaki on which a handful of cattle would graze.

———

A century after the assault on the forests started, the federal govern-ment noted in its first 'Report of the National Estate', in 1974,

235

that the Wet Tropics had been little studied.[3] In 1975, the residual Wet Tropics vegetation communities were mapped by Geoff Tracey and Len Webb in a series of 15 maps. Earlier visits from Webb and Tracey to the forests of Cape Tribulation found, in 1962, that just about every second plant was undescribed. At beautiful Noah Creek between the Daintree River and Cape Tribulation, rare and primitive plants—'green dinosaurs'—were found in 1970 sheltering in secret, shady, moist nooks. Another incidental survey in the mid-1970s discovered 21 new orchid species. When the Forestry Department itself looked at the area between Cape Tribulation and Bloomfield in 1976, they discovered that the rainforest was more extensive than they had realised.[4]

George Quaid wasn't the first to try to ram a road through it. To the first Europeans, who came to the area in the 1930s, the fact that the road along the east coast of Australia was incomplete in just this one place was galling. A craze of forestry tracks had, in 1962, joined up to form a rutted track from the Daintree River to Cape Tribulation. All that was needed, said the settlers, was to join this road with the one from the Bloomfield River, 30 kilometres north, that went on to Cooktown. The government would not pay for such a remote road. So in 1967, the settlers formed a group called the Development League, which included Douglas Shire Council, and attempted to bulldoze it themselves. Bulldozers were brought in from both the north and the south, hoping to meet in the middle, hacking and slashing their destructive way through glorious forests of undescribed species, festooned with orchids, ferns and other epiphytes. The implacable rainforest gazed on as the calamitous machines, like beetle behemoths, became mired in muddy sediment built up over aeons. Utterly trapped, they needed to be hauled out, leaving the rainforest to pick up the pieces and fill the gaps with life.

In 1976, Quaid's reinvigorated efforts were the catalyst for the formation of a local Cape Tribulation environment group. Here was something new. They wanted a large national park at Cape

Tribulation, and they pressured the Forestry Department into assessing the damage being done by Quaid's bulldozers. Trashed trees and rapid gully erosion resulted in a recommendation that work cease immediately.

But in 1983, Joh Bjelke-Petersen—the infamous, uncompromising and corrupt conservative Queensland premier—threw a small amount of money at the project, not enough to carry out an environmental impact assessment, or even a preliminary survey of the route—just enough to let the bulldozers in. When they arrived at the beginning of the track at Cape Tribulation, they were met by a group of wiry and determined protesters who had been working in the background to coordinate a peaceful and non-violent protest. They wanted to draw national attention to the destruction of the rainforest, and the media were on the scene to document the signature police brutality of the time as police tumbled out of vehicles and dragged people into the back of a paddy wagon. The Daintree Blockade had begun.

It was soon discovered that the bulldozer drivers had absolutely no idea where the road was to go. They began to bulldoze along a walking track into the newly gazetted (in 1981) Cape Tribulation National Park. The Queensland National Parks and Wildlife Service had been given clear instructions by the government to facilitate the bulldozing of the road. Senior scientist Peter Stanton, who knew the rainforest better than anyone (and who was to win a national award for drafting a proposal for a series of national parks in the Wet Tropics), found himself between a rock and a hard place. He was left with no option but to walk in front of the dozers, choosing which precious trees should be sacrificed and which should stay, pegging the route of least damage.

People took to the trees. There were heated moments: a chainsaw-wielding council worker attacked protesters; a bulldozer nearly rode over a woman who had fallen out of sight, stopping at the very last moment; a pushed and swaying tree nearly flung a protester who was perched dangerously near the top. As protesters clambered up trees, bulldozers continued to grind dangerously close by. Stanton, who was carefully watching out for protesters and trees alike, was removed from duty because the progress of the dozers was too slow.

The bulldozers continued more freely, and protesters began burying themselves in the ground with just their heads exposed. The images would become international icons. December rain washed the unconsolidated sediment into streams and sent plumes out to the fringing reefs, but still the bulldozing went on. Finally, the wet season of 1983 brought a halt to the works, but in 1984, bulldozing resumed. Police brutality now became more of a feature, and police dogs were released to bring down protesters. Bulldozers continued crashing through trees and even along riverbeds, pushing fine dirt and mud into shocked streams. Protesters began broadcasting to the media from flimsy hammocks strung in treetops. That there were no deaths is a miracle.

In the end, the road was bullied through with sheer bloody-mindedness. It was not the first battle lost by conservationists in the Wet Tropics.

———

In the five or so years leading up to the Daintree Blockade, a small group of young, well-educated, new settlers purchased rainforest-clad land in the Daintree, not to develop it, but to protect it. They comprised an Australian offshoot of the global modern environmental movement that arose post-World War II. Galvanised by the publication of Rachel Carson's *Silent Spring* in 1962 (which alerted the world that industrial society was endangering the environment), it became a social movement in the 1970s that questioned the cost of economic progress and challenged the philosophical underpinnings of consumer society. These new settlers planned to make a go of an alternative lifestyle by building their own homes and growing their own food. But that idyll was shattered almost at inception as members of the group were flung across the forefront of a cascading series of events that would eventually lead to the World Heritage listing of the rainforests of the Wet Tropics.

At nearby Cow Bay, with its pristine low-lying rainforest—one of the last small tracts left after the devastation of King Ranch—a subdivision was surreptitiously being planned. Even though the Douglas Shire Council initially objected to the subdivision, the state government overruled the objection and backed the wealthy real estate developer. A 2700-hectare subdivision was quietly announced by the council and approved without public review in 1981. Almost the first that anyone knew about it was when one of the new settlers stumbled across the developers, who—with a map laid out across the bonnet of a ute—were in the process of splitting the forest into hundreds of small blocks.

With echoes of King Ranch, roads were bulldozed in straight lines without regard to contours and creeks. Boulders were simply blown out of the ground. A policy nightmare to this day, environmental campaigns are still acquiring some of the remaining forest blocks for national parks or, through planning instruments, preventing others from being developed.

At around the same time, the Forestry Department released a plan to log all remaining productive virgin rainforests of the Wet Tropics within five years. The spectacular, isolated, granite massif of the Mount Windsor Tableland was first on the chopping block. With short notice, about a dozen protesters lay down across the road to the Mount Windsor Tableland in a desperate attempt to halt the logging of the richly endemic trees in this significant climate refugia.[5] This proved to be only the second direct-action campaign in the world—the protesters were arrested by police.

The first direct-action campaign was at Terania Creek, in northern New South Wales, in 1979. It was the home of the Bundjalung rainforest people, who looked on with fear while a group of non-Indigenous people were determined to love the forest:[6] it was something unprecedented and terrifying. Terania, at least, was ultimately protected. Mount Windsor Tableland, however, was heavily and carelessly logged—big trees were felled so that they toppled others down to the creek line.[7] The canopy was ripped open, and new roads—costing more than the minuscule royalty earnings from the logs—were gouged across loose slopes, sending smothering blankets of sediment into previously sparkling clean creeks and waterholes filled with endemic freshwater fauna.[8] Unsurprisingly, the blockade of Mount Windsor received a hostile reception from the state government and from the Forestry Department, whose logging practices were now attracting public scrutiny.[9]

Foresters were puzzled by the antagonism. With their new legislation, they had been given an essential place in the rural landscape. And for decades they had battled public apathy, hostility and obstructive government policy in an attempt to extend the area of productive forest. They were instrumental in reserving a large area of national parks. By the time legislation in 1975 divorced National Parks from the Forestry Department, and gave National Parks its own sub-department, there were a million hectares of national parks that attracted two million visitors. Surely the foresters' record as conservationists was without peer.[10] The conservation movement, however, wanted the forests left alone. This was so novel and unexpected that foresters were left perplexed: what do you mean? Do you just want to look at them?[11]

In the early 1980s, two-thirds of Queensland's rainforests were in state forests. The government had not updated its logging quota, which was still based on the outdated and politically driven 1948 quota of 200,000 cubic metres. In 1981, professional foresters doubted that it was possible to obtain even 160,000 cubic metres of good logs. An internal Forestry Department report documenting serious overcutting was suppressed,[12] and sawmills were unable to get enough logs to meet even this lower estimate. In the early 1980s, the quota was officially reduced to 130,000 cubic metres. But only 112,000 cubic metres of timber ready for logging could be found. In 1985, the Forestry Department could find only 88,000 cubic metres to harvest, and by 1987 this was reduced to 60,000 cubic metres.

Employment in the industry plummeted from about 2000 in 1981 to 760 in 1987.[13] It was clear that logging the Wet Tropics had become unsustainable.

The Cairns and Far North Environment Centre, which had been recently constituted, began to coordinate the efforts of the growing number of individuals and groups up and down the coast battling rainforest destruction in their areas, including the Daintree Blockade, and to link these local efforts with the burgeoning national environment movement. In 1982, the Environment Centre joined with the national Australian Conservation Foundation and the state-based Queensland Conservation Council to form the 'Rescue the Rainforests' campaign. This toured the media, politicians and committees, and generated alternative conservation and economic policies such as tourism. (A decade later, the first figures for the value of the Wet Tropics would show total direct spending in the region to be nearly $400 million a year.[14])

But back in the early 1980s, the campaign was dealing with the ruthlessness of a state government that held only contempt for any conservation ideology. In response, the conservation movement—by now strongly networked—bypassed the rancorous state government altogether and launched a national campaign to nominate all of north Queensland's remaining rainforests for World Heritage listing. Quick off the mark, the Australian Conservation Foundation attended the World National Parks Congress in Bali in 1982 to float the idea. The wheels of the international processes began turning when the Wet Tropics was included on the International Union for Conservation of Nature's 1982 indicative list of World Heritage sites,[15] one of the first of such lists.

At the national level, Australian Heritage Commission research was able to confirm the universal natural values of the Wet Tropics in 1984.[16] But the state government would not hear of it. Two years would pass while the federal government dragged its heels waiting for the state government to concede—something that would never happen. The conservation movement simply accelerated its efforts, holding rallies, concerts, events and seminars, and producing articles, books and documentaries. It soon became clear that, at the national level, there was overwhelming community support, even in marginal seats, for go-it-alone federal action that overrode the state. Finally, on the eve of the federal election in 1987, Prime Minister Bob Hawke announced that his government would nominate the Wet Tropics for World Heritage listing if they got back in.

They got back in. And on 19 January 1988, logging in the Wet Tropics was deemed illegal under Australia's *World Heritage Properties Conservation Act 1983*. The Queensland government, predictably, frothed at the mouth. Along with John Howard, the leader of the federal Liberal Party, who was in opposition, they unsuccessfully tried to unravel the listing process by criticising the international evaluation processes and criteria. On 7 December 1988, the Wet Tropics—encompassing nearly 900,000 hectares, along 450 kilometres of coastline—was inscribed on the World Heritage List [17] for its outstanding universal natural values.

Today, the Wet Tropics of Queensland World Heritage Area is regarded as the second most irreplaceable natural site in the world (Canaima National Park in Venezuela is the first)[18] because it features an evolutionary history, from bryophytes to tall flowering rainforests and the glimmerings of the first sclerophyll forests; uninterrupted habitats from antiquity, with their attendant populations of relictual frogs, reptiles and insects echoing Gondwana; globally significant evidence of fauna evolution from marsupials to songbirds; and the mixing of two long-separated floras and faunas when worlds collided at the crumple zone between Australia and Asia. Pleistocene climatic sifting and the links between rainforest and Australia's arid vegetation are all recognised in this exquisitely beautiful place. The unparalleled spectrum of plant communities—classified by science into 250 vegetation types,[19] and managed by rainforest peoples for millennia—represents the richest of Earth's evolution.

Unfortunately for Thiaki, the World Heritage listing would gouge out a new wound.

The World Heritage boundary is a jumble of 620 properties with a small jutting western peninsula, Herberton Range, that skirts around but does not include Thiaki. In 1988, Thiaki sat like an unattached seed, ready to spill from a bursting fruit. In 1992, the funds available for acquisition of properties dried up. By then Barry had had enough of logging and had offered to sell his abused remnant to the government. He threatened to clear the forest if it was not purchased. His insistent phone calls are still remembered to this day in the national park's office. Satellite images taken in 1992 show a claw-shaped gash emerging on very steep slopes in the middle of the remnant as Barry punched his bulldozer in anger against the forest.

# CHAPTER FIFTEEN

# From little things . . .

Rainforest peoples believe that all entities—people, other animals and other beings—are moral agents whose conscious and responsible actions maintain the cosmos. This cosmology, so profoundly different from the European frameworks of power, rejects centralised authority as asocial. Instead, it works to support the status of all entities, none of which has a monopoly on information. Responsible action is based on a slow accumulation of knowledge flourishing into an understanding of complex events. It follows that people are reluctant to intervene in ecological processes except in localised ways or in ways authorised by the accumulation of experience expressed as Dreaming or Law. Under this cosmology, it is impossible to justify the annihilation of groups or species.[1]

———

In a time of upheaval, Ernie Raymont's father quit the farm he was working on, fed up with rations for wages and poor treatment.

It was shortly after the iconic 1966 Wave Hill walk-off, when the Gurindji people of the Northern Territory walked away from their jobs on a cattle station owned by William and Edmund Vestey, unhappy at their poor working conditions and disrespectful treatment. A landmark event, it would bring in—at least conceptually—equal wages and land rights for Aboriginal people. On the Atherton Tablelands, it brought more upheaval. Many people who cleared the land for the settlers and worked the farms were kicked off properties because the owners did not want to pay wages.

While some people got forestry jobs, others—in a breathtaking twist—went looking for gold. Molly, Ernie's grandmother, who had been working in the English family's Malanda Pub, went prospecting. She reclaimed ancestral pathways to pan along the tributaries raking the flanks of Choorechillum (Mount Bartle Frere), drinking from the creeks and living off Country for months. Kids would join the mob during school holidays, and Molly would blow the rain away.[2] The people lived with the spirits of the forest and, for a time, chased gold from the corners of creeks. Perhaps it was a happy time, but it didn't last because declaration of the region as part of the Wet Tropics of Queensland World Heritage Area made prospecting for gold illegal. The people were once again dispossessed.

The fierce and urgent battle to save the rainforest from loggers and developers had resulted in the listing of the Wet Tropics exclusively for its exceptional 'natural' values: the distinction between cultural and natural, so alien to Aboriginal people, was a fundamental plank of European culture. As understandable as the circumstances of nomination and listing were, they nevertheless set up tension between the need to manage the World Heritage Area as part of humanity's universal heritage, and the needs of Aboriginal people who had uniquely occupied and travelled with the forests, shifting with them through Pleistocene climatic vicissitudes and ice ages, volcanic eruptions and anthropogenic climate change.

In 1992, the highest court of law in the land confirmed native-title rights across Australia, including most of the World Heritage

Area. Legislation for the World Heritage Area and the associated Wet Tropics Management Plan were still in development, and the nation-changing High Court ruling came in time to allow several Aboriginal-led initiatives to guide the process. In 1998, a review of Aboriginal involvement in the management of the World Heritage Area led to the 2005 Wet Tropics Regional Agreement, which provided a framework for the cooperative management of the World Heritage Area by rainforest peoples and the government. This was followed, in 2012, by National Heritage listing of the Wet Tropics of Queensland World Heritage Area for its Indigenous heritage values.

Today, nearly 90 per cent of the World Heritage Area is recognised through native-title and Indigenous land-use agreements, including Ngadjon land around Choorechillum. Nearly a third of the World Heritage Area also forms part of the Indigenous Protected Areas.[3] And there are two dedicated Indigenous positions on the Wet Tropics Management Authority board, which is now chaired by an Indigenous woman. It is an unstoppable momentum. On 30 September 2021, the Daintree was returned to the eastern Kuku Yalanji people, who now formally own over 160,000 hectares of rainforest to be jointly managed by the Queensland government. Relisting the World Heritage Area for cultural values is an active area of debate.

If global research on lands managed by Indigenous people is anything to go by, species in the Daintree are now more likely to thrive.[4]

———

Once upon a time, Thiaki was part of a connected rainforest matrix studded with exceptionally rich island jewels[5] spread across the north and down the east side of the continent; they were structured along a continuum of wet to dry, and tropical to temperate habitats. Endurance, survival and persistence are qualities of these rainforests, which are one of Earth's greatest biological and evolutionary success stories.[6]

When we first bought Thiaki, in 2005, it was an island severed by grazing land from its matrix of the Wet Tropics of Queensland World Heritage Area to the south and the Herberton Range National Park to the west. To the north and east are the pastures of the Atherton Tablelands. With binoculars, you can just make out the strings of valiant restoration efforts struggling to tie the remaining precarious lifeboat remnants to each other and to the larger chunks of World Heritage Area. If isolated in perpetual lockdown, even large national parks and World Heritage forests lose resilience, sag and collapse in an extinction vortex.

It is the unimpeded movements of species across whole land-scapes that are as essential to life as the flows of wind and water. Corridors of natural habitat act as superhighways for plants and animals. Birds—pinballs of feathered energy—dart and carouse through them, eating and defecating a rain of seeds. Butterflies and bees flit and bumble through corridors, spreading pollen. Seeds ride zephyrs as if between tall buildings.[7]

Upland ecosystems, like Thiaki's, shift over short distances with greater elevation. Under a warming climate, upland species are driven higher into ever-smaller areas until they approach mountain peaks, where they can go no further. Effectively an escalator to extinction,[8] climate change is driving massive declines in the range of upland birds in the Wet Tropics, with numbers of birds in the lower part of their range crashing by half as birds from lower down surge up. With a 3.5-degree Celsius increase in temperature—anticipated by the end of the century—there are likely to be no areas left in the Wet Tropics with a rich suite of upland species, and all highland endemic vertebrates will probably have abandoned low- and mid-elevation areas.[9]

When patches are all that remain, building resilience across agricultural landscapes by restoring and reconnecting the patches is an immediate conservation priority, not only here but also across the world.[10] Corridors across the sea of agricultural land allow for recolonisation, and immigration to buffer against extinction risk.[11] The umbilical cord to forests also enhances the ecological integrity of agricultural and urban areas, and this integrity is as important as the protected areas themselves.[12]

By 2020, Thiaki had been reconnected, albeit tenuously, to the Wet Tropics of Queensland World Heritage Area to the south, and to the Herberton National Park to the west and north through a series of community plantings. With names such as Lemuroid's Leap and Ringtail Crossing, they are strings of hope and leaps of faith. Often flimsy, plantings as narrow as 20 metres slice across treeless paddocks: lifelines anchoring national parks and remnants on private land to the World Heritage Area. Like refugees in search of a better world, only some animals will cling to these ropy life-lines. The lemuroid ringtail possum, Herbert River ringtail possum, green ringtail possum and Lumholtz's tree-kangaroo, for instance, will travel tentatively and probably only where the corridors are hundreds of metres wide.[13]

A line drawn from the tip of Herberton Range National Park, on a north-easterly heading, approximates the northern edge of the Atherton Tablelands: through Wongabel State Forest, Curtain Fig National Park and the crater lakes of Lake Eacham and Lake Barrine to Gadgarra National Park, it is around 25 kilometres. A slim creek line of trees connects Herberton Range National Park to Wongabel State Forest, from which issues Leslie Creek. Sadly, Leslie Creek is still being cleared for cattle grazing, and in places it is just a scattering of trees that reach skeletal fingers to Curtain Fig National Park. This major tourist attraction is itself lassoed by the community-restored Peterson Creek to the famous crater lakes of Lake Eacham and Lake Barrine, which are tied together by Lake's Corridor and the restoration of Maroobi Creek.[14] Lake Barrine, in turn, is roped to the main World Heritage Area through Donaghy's Corridor, built along spring-fed Toohey Creek that burbles out of Lake Barrine.

In places, Peterson Creek now has the appearance of a well-established forest. But it would take 30 years, and 120,000 planted trees, to establish it along 4.5 kilometres of gullied and gutted creek lines where cassowaries have not been seen since 1975. Restoration of Peterson Creek started in the 1980s at a time when forest clearing and land degradation became a deep public concern after two centuries in which trees were regarded as some sort of enemy to be crushed without compunction.[15] Conservation-minded landholders cobbled together resources to plant small areas at its two ends, at Curtain Fig and Lake Eacham, and around the town of Yungaburra. A trickle of government funding in the late 1990s allowed for fence building and further seedling establishment. Momentum built, albeit slowly, as the community and landholders scraped together bits of cash and applied for government grants, working doggedly despite residual pockets of landholder recalcitrance. In 2011, a tipping point came when a pivotal grazing property was bought by a conservation-minded landholder, who planted Freemans Forest Nature Refuge. Other neighbours, inspired by landholder and community effort, finally saw the benefits of restored creek lines on their own properties and began to fill in the gaps.[16]

Lake Barrine—with its dwindling population of endemic musky rat-kangaroos that are already losing genetic robustness—is close to the greater World Heritage Area of the 80,000-hectare Gadgarra National Park. Shaved as clean of vegetation in the 1930s as a Brazilian wax job, Donaghy's Corridor from Lake Barrine to the national park was restored between 1995 and 1998 when hundreds of people rallied to propagate seedlings and plant over 20,000 trees along a 1.2-kilometre long and 100-metre wide corridor.[17]

But even with goodwill in abundance, there is only so much that can be done without resources and incentives. Today, restoration of the Atherton Tablelands amounts to a couple of thousand hectares spread on tiny holdings embedded in a cleared and often degraded matrix. It is a shocking statistic that three-quarters of the world's remaining forests are now within 1 kilometre of a forest edge.[18]

———

Like the fight to save the rainforests of the Wet Tropics, community momentum on the Atherton Tablelands reflected the global impetus to constrain humanity's war on nature, which reached a crescendo in the 1992 Earth Summit held in Rio de Janeiro, Brazil. Representatives of 178 nations gathered to address the two fused global crises of biodiversity decline and climate change. Ominously, the secretary-general of the United Nations, Boutros Boutros-Ghali, opened the summit with two minutes of silence for life on Earth, dramatically highlighting the urgency of dealing with the two global crises.[19] Two peak international conventions—one for climate change and one for biodiversity—were opened for signature.[20]

Under the 1992 Convention on Climate Change, countries signed up to a target of reducing their emissions of greenhouse gases by five per cent below 1990 levels. They would do this between 2008 and 2012; in the meantime, they would meet to crystallise these commitments in Kyoto in 1997. John Howard, Australia's prime minister in 1997, would once again employ egregious tactics to try to scuttle the momentum of the international climate-change processes, as he had tried with the World Heritage listing of the Wet Tropics. It didn't seem to matter that in our arid and fire-prone country we had the most to lose from climate change. Australia would ensure that it did not need to lift a finger to help save the world.

Australia's economy was steadfastly devoted to the mining industry and to providing cheap coal—cheap because it did not consider the cost of the flow of greenhouse gases into the atmosphere. As well as emitting the highest per capita amount of greenhouse gases, Australia was also bulldozing more forests than any other developed country. Indeed, Australia ranks among the top ten global deforesters, along with Brazil's clearing of the Amazon, Indonesia's clearing of Borneo and the Democratic Republic of Congo, most of the deforestation occurring in Queensland.[21]

Australia went to Kyoto with Machiavellian intent. It threatened to walk away from negotiations if its demands were not met, knowing that this would send a wrecking ball through the fractious Kyoto process. Global negotiators grumbled but caved in to Australia's demand that it shouldn't have to reduce its emissions. It was galling: one of the wealthiest countries in the world was claiming special economic circumstances because it was reliant on coal. Australia not only refused to reduce emissions, but also insisted on being able to actually increase them. In the end, Australia negotiated an increase in its emissions by eight per cent, while the rest of the developed world decreased theirs.

Australia wasn't done. As negotiations wore on through the night, and after exhausted delegates and translators had gone home, Australia picked its moment. It made another astonishing demand: it would only sign up to Kyoto if its 1990 emissions baseline (the year all emissions would be measured against) included emissions from clearing forests. Forest clearing was at its devastating peak in 1990, with skyrocketing emissions. The government knew, however, that Queensland had recently enacted a tough new land-clearing rule, and emissions would plummet. Therefore, including high deforestation emissions in Australia's baseline gave us a free kick—in fact, it gave us a free game. It meant that we would never struggle to meet or beat our Kyoto targets. Pejoratively, this became known as the Australia Clause.

———

And what about biodiversity? With life on Earth in freefall, here was a once-in-a-generation moment to halt its decline by galvanising biodiversity to work hand in hand with energy measures to abate climate change. But, incomprehensibly, biodiversity was ousted (apart from narrowly defined afforestation and reforestation—essentially plantations) from the Kyoto Protocol. In a spectacular own goal, the politics between the European and the US environment movements collided. Europeans—fundamentally reacting to American consumerism—objected to the possibility that the United States could 'buy their way out' of any real effort to reduce their energy emissions by investment in forest conservation and management.

It certainly was true that the United States was wallowing in the extreme free-market ideology ushered in by Ronald Reagan, where selfishness disguised as freedom defined the American way. But, in a farcical theatre, the baby was not only thrown out with the bathwater but also crushed under foot as Europeans sided 'against' forests, and Americans 'for' them. Instead of seeing ecosystems working together with energy abatement in global climate-change deliberations—clearly the original intention—biodiversity became roadkill.

The resulting global policy discord that decoupled biodiversity from climate-change policy has had chaotic ramifications. It has allowed biodiversity to become a convenient global blind spot, and even to be dismissed altogether. Credits from forests were—and still are—banned from the European carbon market because they are not considered 'real' abatement. This is despite European reports showing that avoided deforestation could reduce real extinctions by up to 95 per cent at a carbon price of US$25 per tonne of $CO_2$.[22] In this perverse space, at least 400 million hectares of forest have been cleared since 1990. Assuming around 1000 tonnes of carbon dioxide per hectare, that's equal to 800 years of Australia's emissions at current rates.[23] Then add 25 per cent more for degraded forests. We will never know how much of this could have been avoided by keeping forests and ecosystems in the Kyoto Protocol.

A handful of countries drive global biodiversity decline.[24] Disappointingly, Australia is second on this list of extinction-makers, after Indonesia. Long a climate and conservation laggard, one of the main reasons Australia's biodiversity is decreasing is that the overarching environmental legislation actively facilitates environmental decline.[25] The legislation works on a project-by-project basis for environmental approvals. This allows a steady chipping away at nature since nearly all development projects are approved. An independent review of the national environment legislation (the *Environment Protection and Biodiversity Conservation Act 1999*) could not be clearer about its ineffectiveness. The legislation is singularly unable to halt, let alone reverse, damage that affects threatened species and leads to ecological community decline, or facilitate the restoration of the landscape. Ecosystem decline and the need for restoration are mentioned nearly 300 times in the review.[26]

Because nature is being whittled away incrementally, the damage mostly goes undetected, and that has allowed the federal government to pare back conservation spending: over the past decade, the funding for environment in Australia has been slashed by nearly 40 per cent. Already low in 2013–14, when less than one per cent of Australia's budget was spent on environment and climate combined, the last budget in 2021 allocated one half of one per cent—just $250 million—to address climate change and environment.[27] Such chronic under-resourcing ramifies throughout the length and breadth of the country and from root to branch.

On the Atherton Tablelands, the Commonwealth Scientific and Industrial Research Organisation (CSIRO)—the government's pre-eminent science body—has been undertaking an important survey of tropical forests. It is one of the oldest longitudinal studies in the world, following the forest tree by tree for at least the past 50 years[28] and effectively taking Earth's pulse.[29] There are few long-term studies in the world, even though on-ground measurements are irreplaceable because they check global ecosystem and carbon models and address a thicket of ecological questions in the nexus between biodiversity and climate change that are unable to be answered remotely.

The survey is showing that tropical forests are putting a brake on climate change by sequestering more than 1 billion tonnes of carbon each year—half of the total terrestrial carbon sink. The survey is also showing that, across the world, long-term carbon storage is compromised by loss of the fauna that works with the trees to facilitate ongoing restoration and seed distribution. Forests sequester carbon differentially depending on species, and on the Atherton Tablelands two per cent of the trees—the big trees—account for a third of the carbon in the forest. Surprisingly, Wet Tropics forests store as much as 70 per cent more carbon than rainforests elsewhere; this is probably because, in our arid-adapted country, even the tropical rainforests have become masters of water-use efficiency: where other tropical rainforests might wither during droughts, Australian rainforests thrive and put on more carbon. It is an unexpected and significant finding with important practical applications. Unfortunately, in a country where one billion dollars a year is given to coalmining companies just for diesel-fuel subsidies,[30] funding for the forest survey, which employed one person part-time, has been cut as part of a government lurch away from environment spending.

Over the past 20–30 years on the Atherton Tablelands, private landholders, like us, have bought and converted land to nature refuges. A tally of the 40-plus properties that form the links in one chain that reconnects national parks from the Atherton Tablelands rainforest to the Great Barrier Reef shows an investment of around $18 million. So consequential is the network of community-built corridors[31] that they are now heralded by the Queensland government as of 'State Significance'.[32] Government funding for restoration on the Atherton Tablelands over the same period has amounted to little more than $2 million—less than $100,000 a year, an amount too small to appear in state budgets.

———

In the 1990s, large slabs of Australia were falling into ecological collapse: the Murray-Darling, Australia's most productive river system, was reduced to sludge; poor water quality was threatening the Great Barrier Reef lagoon; and dryland salinity was creeping insidiously across the landscape. Throughout the country, community and landcare groups ramped up local restoration efforts. These actions were impossible for the government to ignore; the Decade of Landcare (1990–2000) surfaced as government policy. Farmers and conservationists enthusiastically aimed to fix land degradation across the country by the year 2000. But it was an impossible feat: the total budget—$340 million over the decade—amounted to less than 5 cents per hectare per year spread across the continent.[33] Even worse, instead of establishing a strategic approach to restoration, the money was made available piecemeal through grants. Groups became mired in the process of applying for grants and worked themselves into exhaustion. When the grants were successful, the focus on restoration was blurred by the need to spend the money quickly.

Meanwhile, across the sweep of the Queensland landscape, clearing accelerated. Prior to the 1990s, the only broad environmental conservation policy applied by the Queensland government in its first century was the soil conservation initiative of the 1960s and 1970s, and this only occurred when it became difficult to ignore the horrific social and environment impact of unsustainable land use.

In 1997, with the increasing levels of public environmental concern, a system to control vegetation clearing on leasehold land began under the *Land Act 1994*, which, for the first time ever, included conservation principles as legislative objectives.[34] Controlling clearing on private land came next, with legislation introduced in 1999 that—with a nod towards Kyoto—also had the objective of reducing greenhouse gases. While the government would use the credits to settle its Kyoto accounts, it did not offer to reward the landholder through stewardship, or ecosystem services payments. Unsurprisingly, landholders reacted with alarm and started deforesting in earnest. Panic clearing reached a peak of 758,000 hectares a year, even after the legislation was in force, and was only slowed with another law that placed a hard moratorium on tree clearing in 2003.

It was obvious that achieving landscape-scale restoration would need more than ad-hoc voluntarism, community involvement and ineffectual legislation. In 1997, the federal government sold part of its telecommunication asset to set up the Natural Heritage Trust. This had a billion-dollar price tag, which bought the beginnings of a more coherent and regionally based approach to natural-resource management. Located in each of Australia's 56 bioregions, the natural-resource management bodies were networked into the community. They were also supported by effective research, with the establishment of Land & Water Australia as well as the Rural Industries Research and Development Corporation. This allowed the bioregions to establish scientifically informed plans to manage region-specific problems. Finally, here was a good idea.

That the approach was working was evidenced by the fact that the network could also scale up solutions. In 2007, for example, the regions of all the catchments flanking the Great Barrier Reef worked with agriculture industries and conservation groups to address declining water quality. The alliance agreed to pollution-reduction targets and catalysed the $200 million Reef Rescue Program.[35]

It all seemed too good to last.

———

I watched as Thiaki blew fractured images into the sky and whispered *nothing is nothing* into the wind. It had stopped raining. On the floor of the house were puddles of water stained with red volcanic mud that marked my progress to the verandah.

Returning my gaze to Thiaki, I decided that I needed to commune with a tree-kangaroo—perhaps irrationally, I had come to see them as a sign of good luck. Slipping and sliding down a pastured slope to the thinly treed creek line below the platypus dam, I moved along the steep eroded embankment with its fuzz of lantana as rain started falling again. I was not wearing a raincoat; water sheeted off my hair and ineffectual clothes. And there, resting on a tree branch, was a tree-kangaroo curled like a palm frond against the deluge. At eye height, as I was on a raised embankment, it turned its head ever so slightly to acknowledge my presence before resuming its antediluvian rumination.

It was February 2007, and I had just returned from coastal Innisfail after a hair-raising drive through the tropical deluge up the Palmerston Highway, winding through World Heritage Wooroonooran National Park and trying not to aquaplane the old ute with its eroded tyre treads around the bends. At Innisfail, I had just signed up for an ecosystem services joint venture, called Degree Celsius, between a company we'd established for the purpose and the Far North Queensland regional body to trial landscape-scale carbon abatement, through payment for ecosystem services. The freshly minted Wet Tropics Regional Plan was to be the framework for the joint venture.

The International Millennium Ecosystem Assessment[36] had just been released and, while it didn't monetise ecosystem benefits, it revealed that most ecosystem services—things that benefit people, such as healthy fisheries, timber, fresh water, soil and pollination services—were deteriorating. Assigning a monetary value, and therefore translating ecosystems into actual policy decisions, was still in the future. Degree Celsius was ahead of its time. And that is a dangerous place to be.

The Wet Tropics Regional Plan signalled restoration as the priority for the Wet Tropics bioregion. Our goal was to incentivise not only this but also the full suite of land-use initiatives, including forest management, clearing avoidance, better management of pasture and beef herds, soil-carbon increases and the reduction of fertiliser use on coastal crops where run-off feeds into the Great Barrier Reef. We aimed to bring ecosystems back into the climate-change picture. If successful, we hoped the model could be rolled out in bioregions throughout Australia. Over a two-year period, we worked up the concept, developing baselines, methods and models of paying land managers through generating carbon credits with biodiversity core benefits.

In 2009, the only way to sell Degree Celsius carbon credits was on the global voluntary carbon market that had begun to flourish after the signing of the Kyoto Protocol. Integrity was essential for land credits, so we arranged to have our project audited against a leading global standard—the Climate, Community and Biodiversity Standards[37]—the only standard in the world that could evaluate a multi-benefit project such as ours. On a high note, we enthusiastically flew in auditors from overseas.

Playing in the background, however, was a discordant tune. Unfortunately, and as often happens, the new federal government was intent on differentiating itself from its predecessor. Kevin Rudd's Labor government decided to abort the regional model that was showing so much promise, and slashed regional budgets by nearly half. Funds were instead diverted to other priorities, and the government was back to awarding grants competitively to small groups: those who had just been working collaboratively to help manage the complex problems of a century of unsustainable land use were once again competing for a dwindling bucket of money. Degree Celsius was down. But it was not yet out.

The government had boldly launched an emissions-trading scheme, the Carbon Pollution Reduction Scheme. It was under attack by both industry, which did not want to pay for emissions reduction, and the conservation sector, which wanted the scheme to be even tougher on industry. Knowing the chances of getting the emissions-trading scheme through parliament were slim, the government was in furious negotiation with the opposition on amendments that would sweeten its passage. The scheme had taken its cue from the Kyoto Protocol and had virtually ignored the ecosystem sector. We were able to lobby during this tumult for amendments that allowed the crediting of the full suite of land-sector abatement initiatives.

As it turned out, the scheme was shambolically abandoned anyway by Prime Minister Rudd, and this triggered his ousting in

June 2010. A new emissions-trading scheme was launched by the next prime minister, Julia Gillard. Enacted in 2011, this scheme carried with it an enhanced land-sector package, including the full suite of land-sector initiatives now rebadged as the Carbon Farming Initiative. It sat alongside the emissions-trading scheme to become the first national offset scheme in the world to include carbon credits derived broadly from ecosystems. Gillard lost the next election to Tony Abbott, a climate sceptic. As prime minister, his first act was to repeal the emissions-trading scheme. The Carbon Farming Initiative, however, survived and forms the foundation of today's national climate scheme that, while voluntary, could easily evolve into an emissions-trading scheme that worked hand in hand with ecosystems to abate climate change and biodiversity decline.

In the end, six prime ministers were to be deposed in Australia's ongoing climate wars,[38] including the recently defeated Prime Minister Scott Morrison, who ridiculed climate action in his 2022 election campaign. A new Labor government led by Prime Minister Anthony Albanese was elected with a strong mandate to end the climate wars.

———

While Australia got what it wanted in 1997 at Kyoto, it disdainfully refused to uphold its end of the deal. Another decade would pass before Prime Minister Kevin Rudd would sign the Kyoto Protocol as the first act of his government.

With targets set, the Kyoto Protocol did indeed drive energy transformation globally and, in Australia, grudgingly. So that Australia would not look like a complete climate pariah,[39] a pitiable Renewable Energy Target of two per cent was agreed in 1997 and a Mandatory Renewable Energy Target was introduced in 2001 known as the two per cent MRET.[40] Surprisingly, however, this was enough to catalyse renewable energy research and deployment. In 2009, under Rudd, it was raised to 20 per cent, and was supported by a $10 billion financing agency for clean energy, and an Australian Renewable Energy Agency to fund clean-tech research including the export of wind and solar power. Even a change in government back to climate scepticism under Tony Abbott could not stop the momentum as big investors slid inexorably, like melting glaciers, away from coal. In 2020, nearly 30 per cent of Australia's total electricity generation came from renewable energy sources.

Internationally, innovation through international patents increased immediately after the Kyoto Protocol was adopted in both developed and developing countries, showing the catalytic importance of international policy for diffusion of technology.[41] Across Europe, in particular, a suite of integrated policies established affordable, secure and sustainable energy technologies, while innovations and industries are driving the efficient transition to net-zero emissions societies. Financial programs totalling over €100 billion underpin innovation and research.[42]

———

In the meantime, as investment in renewable energy soared, investment in biodiversity and nature-based solutions to climate change never got off the ground. Blind to the fact that humanity is embedded within nature, global business continued pursuing an infinite growth and development agenda that amplified environment degradation using the standard-bearing measure of gross domestic product (GDP). GDP signally ignores the erosion, or depreciation, of the natural capital on which it is based, even though most of the world's GDP is dependent on nature, and most businesses on the planet are exposed one way or another to risk associated with loss of nature.[43]

As recently as the late twentieth century, even the idea that economists and ecologists should work together was novel.[44] In 1997, a group of ecological economists boldly made a monetary estimate of the value of the global flow of Earth's services to humans. Their estimates of US$16–54 trillion annually were larger than global GDP at the time.[45] Criticisms that such absolute values are meaningless (not the least because if life was destroyed, there would be no one to receive this wealth[46]) missed the point of the exercise, which was to show the value of biodiversity in order to influence policy decisions.

The Intergovernmental Panel on Climate Change (IPCC), which recently announced unprecedented changes to Earth's climate,[47] has certainly been successful in galvanising policy. Formed in 1988, it catalysed the Kyoto Protocol. The IPCC's biodiversity sibling, the Intergovernmental Science-Policy Platform on Biodiversity and Ecosystem Services (IPBES), was formed in 2012 but only gained real attention in 2019 with a shocking assessment that a million species were threatened with extinction; that nearly 96 per cent of mammal biomass on the planet comprises humans and their livestock; and that nearly 80 per cent of land and most of the ocean have been modified by human activities.[48] The biodiversity platform found that the decline in ecosystem services was having a grave impact on people the world over.[49] Only 40 per cent of the world's remaining forests now have a high ecosystem integrity, little modified by human activity.[50] The Convention on Climate Change remains slow to formally accept the interconnectedness of climate change and biodiversity decline. While never stated, it seems to me that the general view, while incorrect, is that if climate change is managed, so is biodiversity decline.

In 2015, biodiversity's clarion call to action finally came with the formal signing of the Paris Agreement (the twenty-first Conference of the Parties under the Convention on Climate Change). This stipulated that global temperature increase should be limited to 1.5 degrees Celsius[51]—but it was simply not possible to do this without harnessing ecosystems, restoration and nature-based solutions.[52] By 2021, two-thirds of the world's countries had responded to the call, with nature-based solutions to climate change as part of their national climate commitments. According to UN Secretary-General António Guterres, 'drastically reducing deforestation and systemically restoring forests and other eco-systems is the single largest nature-based opportunity for climate mitigation'.[53]

A barrage of research arguably instigated by the Paris Agree-ment predicts that managing forests, avoiding land clearing, better managing pasture and beef herds, harnessing Indigenous fire management, increasing soil carbon and reducing fertiliser use will provide nearly 40 per cent of the cost-effective mitigation needed in the next decade to hold climate change below 2 degrees Celsius.[54] In fact, global modelling is showing that the planet's future climate is intimately tied to the future of its forests, which not only remove a quarter of the $CO_2$ that humans emit, but also hold more carbon than exploitable oil, gas and coal, and create rainfall for growing food.[55] The Glasgow Climate Pact[56] has cemented the 1.5-degree-Celsius ambition and demanded rapid, deep and sustained emissions reduc-tions: 45 per cent by 2030, and net zero by mid-century, including by 'protecting, conserving and restoring nature and ecosystems'.[57]

The World Economic Forum has newly instigated the Nature Action Agenda,[58] which is aimed at catalysing economic action to halt biodiversity loss by 2030. The economic urgency is under-scored by a recent analysis, by Her Majesty's Treasury in the United Kingdom, which shows that delaying action to stabilise biodiversity intactness by even a decade will mean that the cost will double, and it may simply become unfeasible to stabilise biodiversity to even the depleted levels of today.[59]

Swiss Re, one of the world's leading providers of insurance, has also developed a Biodiversity and Ecosystem Services risk index that is being incorporated into their decision-making. For Australia, it's not looking good. Swiss Re has placed it near the bottom of the world's countries for decline in ecosystem services requiring restoration,[60] since Australia's environment policy settings make it a high-risk country for investment. Australia's gamble on future tech-nology leaves biodiversity a bloodless abstraction that the country can seemingly get along without.[61] It seems we have learned very little since the policy of closer settlement, which destroyed the rainforests and the livelihoods of rainforest peoples and sent settlers on a journey of destitution and heartache.

The first blueprint to tackle the nexus of climate, biodiversity and humans was also released only in 2021.[62] It requires nothing less than the transformation of humanity's relationship with nature: integrating conservation and restoration across the landscape; reforming national economic, legal and financial systems so that they account for nature; and establishing carbon pricing and payment for ecosystem services—transforming governance structures, financial institutions and businesses both big and small.

The concept of nature-based solutions to climate change is now resonating across the globe. Ironically, while Australia's politicians scrapped in the grotty sandpit of the climate wars, the Carbon Farming Initiative has quietly become the largest and arguably most respected natural systems carbon market in the world, albeit only in the voluntary market. Over the years, it has been refined and strengthened, anchored with science based systems, and underpinned by strict regulation to ensure integrity. It trades ecosystem services across the landscape, including from trees, farms, soil and fire: Indigenous fire management across the 2 million square kilometres of Australian savannas—that fine-scale nurturing of Country—now benefits people to the tune of a quarter of a billion dollars per annum, while abating over 7 million tonnes of $CO_2$.[63] The Australian Carbon Credit Unit is now one of the most highly regarded carbon credits in the world. In Queensland, it has catalysed a half-billion-dollar land-restoration fund to measure and trade core environmental and social benefits, not just carbon.[64]

The challenge for the future is to harness big data and computer algorithms—refined geospatial information, image processing, machine learning and sophisticated statistical testing—to enable the changes in carbon and biodiversity from land-management activities across landscapes to be measured. It may be possible to name every tree—including my Tree—in the not-too-distant future. Like the shift to renewable energy, the shift towards a nature-positive economy is picking up speed.

Suddenly, coal and gas—and indeed any corporations that are shown to destroy nature—have become high-risk investments. Has the world suddenly lurched onto a different track? A wall, not of water or billowing asteroid-blasted dust, but of money is making its way towards nature. And I completely relate to the Terania Creek people, who were frightened by the sudden love.

———

In mid-2020, another young cassowary strolled across our restored valley. Carrying the weight of community restoration efforts on its feathered brown back, I hoped it would travel to the growing network of linking restoration corridors, rather than the intersecting 100-kilometre-per-hour highways—or the neighbour's place.

I punch a hole in the ground with the planter spade—a neat incision, more like keyhole surgery than open-cut—reach into the bag of seedlings hanging from my hips and then place a small seedling into the slot. A gentle nudge with the toe of my boot sutures the hole shut. Professional planters can plant up to 1500 trees a day on the steep slopes of Thiaki. I can get in around 60 an hour if I don't pause to admire each seedling. Each is a bright green packet of potential, a little force of nature and a bloom of hope. And each has its story to tell. My pouch of seedlings carries a world of stories that reflect the diversity of Thiaki, a composite of evolving species with Gondwanan and Asian heritage that, while conservative, have on occasion stepped out to evolve into the adjacent open forest environments.

*Flindersia brayleyana* (Queensland maple)—which we planted as a monoculture—is endemic to north-east Queensland. I squint to see the abundant oil dots in the seedling's elliptical leaves. As a tree, this seedling will produce pendulous capsular fruits that open in a starburst, out of which will float the winged seeds. *Flindersia's* family is Rutaceae. There are fossils of this family from Gondwanan times. Most of the 17 *Flindersia* species are rainforest plants, but at least one—*F. maculosa* (leopardwood)—has broken with tradition and headed into arid areas. *Castanospora alphandii* (brown tamarind), another endemic, has veins that loop inside the blade margin. Its family, Sapindaceae, is related to Rutaceae; both are widespread globally, with a strong Gondwanan presence. The two families presumably made their way to Australia across Antarctica. *C. alphandii* will grow into a tall dense-crowned tree with peach-coloured fruits that are loved by cassowaries.

In addition to *Flindersia* and *Castanospora*, our six-species mix includes *Cardwellia sublimis* (northern silky oak), *Ficus septica* (septic fig), *Neolitsea dealbata* (hairy-leaved bolly gum) and *Syzygium cormiflorum* (bumpy satinash). *C. sublimis* is another species endemic to the Wet Tropics. On the forest floor, its seedlings resemble a kaleidoscope of green-winged butterflies; some of them will become large emergent trees with delicately scented cream flowers. The species evolved perhaps 30 million years ago from a Proteaceae ancestor whose line goes back to the Cretaceous over 100 million years ago.

I caress the orbicular leaves of *Ficus septica*; they are gently hairy underneath. This fig prefers disturbed areas, and it is widespread in south-east Asia and the Pacific where it is eaten by parrots, doves, orioles, cuscuses and at least 13 different types of bats.[65] The fruits are pollinated by fig wasps.

Featuring twigs clothed in erect tortuous hairs, *Neolitsea dealbata* is wildly attractive to the male tooth-billed bowerbird; he fastidiously turns individual leaves in his cleared display court so that the white undersides are shown, in the hope of attracting a female. The species is dioecious (with separate male and female plants). While it is endemic to Australia, its ancestors came here millions of years ago from Asia. Its family, Lauraceae, has perhaps 3000 species, and their fossils have been found imprinted in Cretaceous rock worldwide.

*Syzygium cormiflorum* has oval-shaped leaves that are drawn out to a tip, and the veins loop strongly inside the margin of the leaves. As a tree, this seedling will bear its white apple-like fruits as eruptions from bumps along the trunk. *Syzygium* is an Australian genus in the Myrtaceae family, which evolved in Gondwana perhaps 60 million years ago. *Syzygium* itself evolved in the Australian rainforests about 30 million years ago. Birds with bellies full of *Syzygium* fruits flew to Asia, making *Syzygium* one of the most successful immigrants to south-east Asia.[66] The genus spread abundantly across the rainforests of Indonesia and into mainland Asia, evolving into at least 1200 species.

*Lomatia* is a member of the Proteaceae family, and there are fossils of it from the Palaeocene about 60 million years ago. Recently, its flowers have been found to have the highest cyanide concentration of any plant tissue[67]—which will make me think twice before dipping my tongue in its nectar-gleaming flowers. Why *Lomatia* should have such toxic flowers—whether as a deterrent to herbivores or as a by-product of metabolism—remains a secret known only to *Lomatia* for now. I try to speed up my planting without stopping to ponder each plant's past and future.

From above, our paddock is now a pattern of plantings that echoes the chequered motif of a Ngadjon rainforest shield, daubed with restoration designs. Thiaki denizens are already inserting themselves, with splashes, dots and corridors of new recruits creating unique and varied versions of themselves. In time it will be difficult to see the original planted trees. Some plots tucked against the forest have already merged. Other plots further away are beginning to assemble into self-sustaining closed canopies. There have been a few areas where the plantings have struggled, mostly reflecting the conditions at the time of planting. We will continue to tease out the lessons Thiaki has set for us.

Using common-sense approaches—planting when the ground is wet; utilising small tubestock plants, wider seedling placement, judicious species mixes, and quick and efficient insertion with minimal disturbance by planter spade; and spraying in rows rather than blitzing entire slopes with expensive herbicide—we have shown that it is possible to build a rainforest from a paddock for as little as $5000–$8000 per hectare. At these prices, it is cost effective to restore the 54,000 hectares of the Wet Tropics that need it on steep slopes, or on land of poor agricultural quality or shredded of vegetation along waterways.[68]

The big picture of our restoration efforts has aimed to optimise carbon sequestration, biodiversity recovery and livelihood benefits. It seems as if the tide has indeed turned for ecosystem services: through the Carbon Farming Initiative, we have now made an income, attracting a premium from our biodiverse-rich carbon credits that easily matches anything we could earn from cattle grazing alone.[69] Our 20 cattle grazing on richer flats are now the supplementary income supporting carbon and biodiversity.

It is not lost on me that we have done this in the same year (2021) that the first ever collaboration between the climate change and biodiversity conventions was held.[70] Though I am at a loss to understand why it has taken quite so long for non-rainforest people to see that climate, biodiversity and humans are one intimately linked system. Seen through this prism, it becomes clear that ecological function rests across whole landscapes in climate refugia, corridors, shifting climate belts and other locally based working landscapes.

I look up to the surrounding World Heritage forest in which Thiaki sits like a seed in a flower carpel. Thiaki's high places, tracking Gondwana, continue to contribute to the rainforest assembly across the southern half the world,[71] while its flanks are dotted with Asian invaders. Red cedar is one of these, its small winged seeds parachuting down during glacial-strength westerlies one or two million years ago, and rapidly infiltrating to spread throughout the fertile high-light environments of eastern Australia. Other species are waiting in the shadows to emerge when conditions change once again.[72]

Ecosystems don't move as a unit but as individuals proceeding at their own pace according to their own whims. This may have been obvious to the rainforest peoples who were themselves stitched into the patterns of the forest. Deep in their fabric, people would have known that chaotic swings in climate lead ecosystems to literally disassemble and reassemble into novel new ecosystems.[73] Our landscape is an interlocking creative system ranging across forever time and microbial time. This infinitely intricate system is always shifting and evolving, reacting and adapting to driving disturbances, always self-organising towards complexity, diversity and resilience. Thiaki will never be able to return to a landscape that existed before European invasion, but 300 million years of living dangerously has given the forest an interesting perspective on the world . . . and a prescience displayed in the miracle of regeneration.

# ACKNOWLEDGEMENTS

First and foremost, I acknowledge Thiaki and the rainforest peoples who nourished and nurtured the forest over millennia. Traditional Owners Ernie and Margaret Raymont were gracious and welcoming during my frequent visits to discuss their history and that of the rainforest. Ernie keenly read my drafts about his story and that of his countryman, Lumholtz's tree-kangaroo, fascinated by the scientific take on events. Yvonne Canendo checked the story of the formation of the Yamani and Barriny. Barry and Kerstin Pember sold us Thiaki.

I wanted to write a narrative from the point of view of the forest, or at least of events that would most mould the forest's story. I sent a teaser to an old friend and mentor Peter Mitchell, and to Tom Griffiths, and it was their encouragement that kept me going. Sean Ulm helped open my mind to those extraordinary rainforest peoples—those future makers, whose creative cosmology sustained the lands that comprise Australia.

My thanks to Renier van Raders and Petrina Callaghan of the Eacham Historical Society for generously responding to my queries.

Scott Hocknull and Mike Archer delightfully conveyed the excitement of prehistoric discovery. Keith Smith was always available to help with issues concerning nature refuges. Stuart Worboys assisted with plant identification and showed me the magnificent *Stockwellia quadrifida* stand of trees. Stuart also introduced me to QuestaGame, a mobile app game for identifying flora and fauna which occupied my family and sent the three generations of us—Noel and I, son Luke, and joyous grandchildren Milla and Dan—into a competitive frenzy during the 2020 lockdown at Thiaki (that Luke won is still in dispute). We learned to examine every angle of leaf, trunk, root, soil and sky for life hidden in plain sight.

My thanks also to Michael Bird, Robin Beck, Mike Berwick, Matt Bradford, Dave Cassells, Wendy Cooper, Darren Crayn, Dave Gillieson, Asa Ferrier, Susan Laurance, Roger Martin (who alerted me to the population expansion and contraction of Bennett's tree-kangaroo and may not like my interpretation), Geoff Monteith, Mike and Helen Murdoch, Al Pember, Dave Rentz, Roy Rickson, Peter Stanton, Chris Tsilemanis and Sam and Martin Willis.

My thanks also to Terri-ann White, now of Upswell Publishing, who originally offered to publish the book through UWA Press, and then fell on difficult times. The fact of the offer kept me going for most of the writing. Heartfelt thanks go to Elizabeth Weiss of Allen & Unwin, who took the book on and thought it had legs. And to Dannielle Viera and Angela Handley for the meticulous editing.

My husband, Noel, was, as ever, supportive, patient and loving.

In 2021 three young Yindji men visited Thiaki to look at our restoration efforts. Their life stories, present activities and future aspirations are an inspiration. So thank you Rangi, Jai, Daniel and dynamo of a sister Kylie Clubb for that much-needed burst of hope.

# NOTES

## Chapter One: Ambush

1   Homer. *The Odyssey*. Translated by E.V. Rieu. Revised by his son D.C.H. Rieu. 1991. Penguin Books.
2   Homer. *The Odyssey*.
3   Berndt, C.H. and R.M. Berndt. 1983. *The Aboriginal Australians: The first pioneers*. Pitman.
4   Berndt, C.H. and R.M. Berndt. 1983. *The Aboriginal Australians*.
5   Rose, D.B. 2013. Death and grief in a world of kin. pp. 137–47 in G. Harvey, editor. *The Handbook of Contemporary Animism*. Routledge.
6   Nicholls, C.J. 2014. 'Dreamings' and dreaming narratives: What's the relationship? https://theconversation.com/dreamings-and-dreaming-narratives-whats-the-relationship-20837.
7   Rose, D.B. 1996. *Nourishing Terrains: Australian Aboriginal views of landscape and wilderness*. Australian Heritage Commission.
8   Sutton, P. 1978. *Wik: Aboriginal society, territory and language at Cape Keerweer, Cape York Peninsula, Australia*. University of Queensland.
9   Rose, D. 2005. An Indigenous philosophical ecology: Situating the human. *The Australian Journal of Anthropology* 16:294–305.
10  Sutton, P. 1978. *Wik*.
11  Gammage, B. 2012. *The Biggest Estate on Earth: How Aborigines made Australia*. Allen & Unwin. p. 133.
12  Docker, E.G. 1965. *Simply Human Beings*. Angus & Robertson.
13  Mountford, C.P. 1976. *Nomads of the Australian Desert*. Rigby.
14  Rose, D.B. 2000. To Dance with Time: A Victoria River Aboriginal study. *The Australian Journal of Anthropology* 11:287–96.

271

15    Idriess, I. 1994. *Back o' Cairns*. HarperCollins. p. 285.
16    Gammage, B. 2012. *The Biggest Estate on Earth*. p. 133.


**Chapter Two: The getting of wisdom**

1    Frawley, K.J. 1983. Forest and land management in north-east Queensland: 1859–1960. PhD thesis. ANU.
2    Winter, J.W., F.C. Bell, L.I. Pahl and R.G. Atherton. 1987. Rainforest clearing in northeastern Australia. *Proceedings of the Royal Society of Queensland* **98**:41–57.
3    McJannet, D., J. Wallace and P. Reddell. 2007. Precipitation interception in Australian tropical rainforests: II Altitudinal gradients of cloud interception, stemflow, throughfall and interception. *Hydrological Processes* **21**:1703–18.
4    Kenrick, P. and P.R. Crane. 1997. The origin and early evolution of plants on land. *Nature* **389**:33–9.
5    Kenrick, P. and P.R. Crane. 1997. The origin and early evolution of plants on land.
6    Bond, D.P.G. and S.E. Grasby. 2017. On the causes of mass extinctions. *Palaeogeography, Palaeoclimatology, Palaeoecology* **478**:3–29.
7    Calvo, P., V.P. Sahi and A. Trewavas. 2017. Are plants sentient? *Plant, Cell & Environment* **40**:2858–69.
8    Tims, J.D. and T.C. Chambers. 1984. Rhyniophytina and Trimerophytina from the early land flora of Victoria, Australia. *Palaeontology* **27 Part 2**:265–79.
9    https://museumsvictoria.com.au/melbournemuseum/resources/600-million-years/ (accessed 1 August 2022).
10    Deleuze, G. and F. Guattari. 1987. *A Thousand Plateaus*. University of Minnesota Press.
11    Dahl, T.W., E.U. Hammarlund, A.D. Anbar et al. 2010. Devonian rise in atmospheric oxygen correlated to the radiations of terrestrial plants and large predatory fish. *Proceedings of the National Academy of Sciences* **107**:17911–15.
12    Beerling, D.J., E.P. Kantzas, M.R. Lomas et al. 2020. Potential for large-scale $CO_2$ removal via enhanced rock weathering with croplands. *Nature* **583**:242–8.
13    Strullu-Derrien, C., M.-A. Selosse, P. Kenrick and F.M. Martin. 2018. The origin and evolution of mycorrhizal symbioses: From palaeomycology to phylogenomics. *New Phytologist* **220**:1012–30.
14    Cheesman, A.W., N.D. Preece, P. van Oosterzee et al. 2018. The role of topography and plant functional traits in determining tropical reforestation success. *Journal of Applied Ecology* **55**:1029–39.
15    Skene, K.R. 2003. The evolution of physiology and development in the cluster root: Teaching an old dog new tricks? *Plant and Soil* **248**:21–30.
16    Dahl, T.W., E.U. Hammarlund, A.D. Anbar et al. 2010. Devonian rise in atmospheric oxygen correlated to the radiations of terrestrial plants and large predatory fish.
17    https://en.wikipedia.org/wiki/Evolution_of_tetrapods#cite_note-Hilderbran-23 (accessed 1 July 2019).
18    https://www.abc.net.au/science/ozfossil/ageofreptiles/fauna/labyrinthodont.htm (accessed 1 July 2019).
19    Weinert, L.A., J.H. Werren, A. Aebi et al. 2009. Evolution and diversity of Rickettsia bacteria. *BMC Biology* **7**:6.
20    Barker, S.C., A.R. Walker and D. Campelo. 2014. A list of the 70 species of Australian ticks; Diagnostic guides to and species accounts of *Ixodes holocyclus* (paralysis tick), *Ixodes cornuatus* (southern paralysis tick) and *Rhipicephalus australis* (Australian

cattle tick); and consideration of the place of Australia in the evolution of ticks with comments on four controversial ideas. *International Journal for Parasitology* 44:941–53.

21  White, M.E. 2006. Environments of the geological past. pp. 17 50 in J.R. Merrick, M. Archer, G.M. Hickey and M.S.Y. Lee, editors. *Evolution and Biogeography of Australasian Vertebrates*. Auscipub.

22  Bond, D.P.G. and S.E. Grasby. 2017. On the causes of mass extinctions.

23  Johnson, D. 2009. *The Geology of Australia*. Second edition. Cambridge University Press.

24  Bond, D.P.G. and S.E. Grasby. 2017. On the causes of mass extinctions.

25  Tewari, R., S. Chatterjee, D. Agnihotri and S.K. Pandita. 2015. *Glossopteris* flora in the Permian Weller Formation of Allan Hills, South Victoria Land, Antarctica: Implications for paleogeography, paleoclimatology, and biostratigraphic correlation. *Gondwana Research* 28:905–32.

26  Goosem, S. 2002. Update of Original Wet Tropics of Queensland Nomination Dossier. Wet Tropics Management Authority.

27  https://en.wikipedia.org/wiki/Terra_Nova_Expedition#Scientific_legacy (accessed 10 June 2020).

28  Pigg, K.B. and S. McLoughlin. 1997. Anatomically preserved *Glossopteris* leaves from the Bowen and Sydney basins, Australia. *Review of Palaeobotany and Palynology* 97:339–59.

29  van Oosterzee, P. 1997. *Where Worlds Collide: The Wallace Line*. Reed.

30  Brenner, E.D., D.W. Stevenson and R.W. Twigg. 2003. Cycads: Evolutionary innovations and the role of plant-derived neurotoxins. *Trends in Plant Science* 8:446–52.

31  Brenner, E.D., D.W. Stevenson and R.W. Twigg. 2003. Cycads.

32  Attenborough, D. 1979. *Life on Earth*. Collins.

33  Rivadeneyra-Domínguez, E. and J.F. Rodríguez-Landa. 2014. Cycads and their association with certain neurodegenerative diseases. *Neurología* 29:517–22.

34  https://www.youtube.com/watch?v=w6YX8AWXrCA (accessed 10 June 2020).

35  Reisz, R.R. and J. Fröbisch. 2014. The oldest caseid synapsid from the Late Pennsylvanian of Kansas, and the evolution of herbivory in terrestrial vertebrates. *PLoS One* 9:e94518.

36  Salgado, L., J.I. Canudo, A.C. Garrido et al. 2017. A new primitive Neornithischian dinosaur from the Jurassic of Patagonia with gut contents. *Scientific Reports* 7:42778.

37  https://anthropologyfromtheshed.com/project/the-ancient-practice-of-macrozamia-pit-processing-in-southwestern-australia/ (accessed 15 June 2019).

38  McConnel, U. 1931. A moon legend from the Bloomfield River, North Queensland. *Oceania* 2:9–25.

39  Banks, J. 2005. *The Endeavour Journal of Sir Joseph Banks (Journal from 25 August 1768–12 July 1771)*. Project Gutenberg Australia. http://gutenberg.net.au/ebooks05/0501141h.html#apr1770.

40  Preparation of cycads from Roth, W.E. 1901. *North Queensland Ethnography*. Government Printer, Brisbane. Yalanji-Warranga Kaban. 2004. *Yalanji People of the Rainforest Fire Management Book*. Little Ramsay Press.

## Chapter Three: The new world

1  Santibáñez, P., A.M. Palomar, A. Portillo et al. 2015. The role of chiggers as human pathogens. pp. 173–202 in A. Samie, editor. *An Overview of Tropical Diseases*. IntechOpen. https://www.intechopen.com/books/an-overview-of-tropical-diseases.

2   Abliz, O., S. Huimin and V. Squires. 2014. Mites (Acari): Small but significant ecosystem components. UNESCO–EOLSS. https://www.academia.edu/7486112/ Mites_Small_but_significant_ecosystem_components.

3   Thulborn, T., A. Warren, S. Turner and T. Hamley. 1996. Early Carboniferous tetrapods in Australia. *Nature* **381**:777–80.

4   Sellwood, B.W. and P.J. Valdes. 2006. Mesozoic climates: General circulation models and the rock record. *Sedimentary Geology* **190**:269–87.

5   Leslie, A.B., J. Beaulieu, G. Holman et al. 2018. An overview of extant conifer evolution from the perspective of the fossil record. *American Journal of Botany* **105**:1531–44.

6   Calvo, P., V.P. Sahi and A. Trewavas. 2017. Are plants sentient? *Plant, Cell & Environment* **40**:2858–69.

7   Brodribb, T.J., J. Pittermann and D.A. Coomes. 2012. Elegance versus speed: Examining the competition between conifer and angiosperm trees. *International Journal of Plant Sciences* **173**:673–94.

8   Brodribb, T.J., J. Pittermann and D.A. Coomes. 2012. Elegance versus speed.

9   de Keyser, F. and K.G. Lucas. 1968. *Geology of the Hodgkinson and Laura Basins, North Queensland.* Bureau of Mineral Resources, Geology and Geophysics, Commonwealth of Australia.

10   de Keyser, F. and K.G. Lucas. 1968. *Geology of the Hodgkinson and Laura Basins, North Queensland.*

11   Labandeira, C.C. and J.J. Sepkoski. 1993. Insect diversity in the fossil record. *Science* **261**:310–15.

12   Simões, T.R., M.W. Caldwell, M. Tałanda et al. 2018. The origin of squamates revealed by a Middle Triassic lizard from the Italian Alps. *Nature* **557**:706–9.

13   Cisneros, J.C., C. Marsicano, K.D. Angielczyk et al. 2015. New Permian fauna from tropical Gondwana. *Nature Communications* **6**:8676.

14   Jasinoski, S.C. and F. Abdala. 2017. Aggregations and parental care in the Early Triassic basal cynodonts *Galesaurus planiceps* and *Thrinaxodon liorhinus*. *PeerJ* **5**:e2875.

15   Bond, D.P.G. and S.E. Grasby. 2017. On the causes of mass extinctions. *Palaeogeography, Palaeoclimatology, Palaeoecology* **478**:3–29.

16   Labandeira, C.C. and J.J. Sepkoski Jr. 1993. Insect diversity in the fossil record. *Science* **261**:310–15.

17   Retallack, G.J. 2013. Permian and Triassic greenhouse crises. *Gondwana Research* **24**:90–103.

18   Sobolev, S.V., A.V. Sobolev, D.V. Kuzmin et al. 2011. Linking mantle plumes, large igneous provinces and environmental catastrophes. *Nature* **477**:312–16.

19   Retallack, G.J. 2013. Permian and Triassic greenhouse crises.

20   Bond, D.P.G. and S.E. Grasby. 2017. On the causes of mass extinctions.

21   Diaz, R.J. and R. Rosenberg. 2008. Spreading dead zones and consequences for marine ecosystems. *Science* **321**:926–9.

22   Bond, D.P.G. and S.E. Grasby. 2017. On the causes of mass extinctions.

23   Smith, R.M.H. and J. Botha-Brink. 2014. Anatomy of a mass extinction: Sedimentological and taphonomic evidence for drought-induced die-offs at the Permo-Triassic boundary in the main Karoo Basin, South Africa. *Palaeogeography, Palaeoclimatology, Palaeoecology* **396**:99–118.

24   Chen, Z.-Q. and M.J. Benton. 2012. The timing and pattern of biotic recovery following the end-Permian mass extinction. *Nature Geoscience* **5**:375–83.

25   Nowak, H., E. Schneebeli-Hermann and E. Kustatscher. 2019. No mass extinction for land plants at the Permian–Triassic transition. *Nature Communications* **10**:384.

26   Fielding, C.R., T.D. Frank, S. McLoughlin et al. 2019. Age and pattern of the southern high-latitude continental end-Permian extinction constrained by multiproxy analysis. *Nature Communications* **10**:385.

27   Warren, A.A. 1980. *Parotosuchus* from the Early Triassic of Queensland and Western Australia. *Alcheringa: An Australasian Journal of Palaeontology* **4**:25–36.

28   McLoughlin, S. 2001. The breakup history of Gondwana and its impact on pre-Cenozoic floristic provincialism. *Australian Journal of Botany* **49**:271–300.

29   Chen, Z.-Q. and M.J. Benton. 2012. The timing and pattern of biotic recovery following the end-Permian mass extinction.

30   Niedźwiedzki, G., P. Bajdek, K. Owocki and B.P. Kear. 2016. An Early Triassic polar predator ecosystem revealed by vertebrate coprolites from the Bulgo Sandstone (Sydney Basin) of southeastern Australia. *Palaeogeography, Palaeoclimatology, Palaeoecology* **464**:5–15.

31   Jablonski, D. 2005. Mass extinctions and macroevolution. *Paleobiology* **31**:192–210.

32   Payne, J.L. and M.E. Clapham. 2012. End-Permian mass extinction in the oceans: An ancient analog for the twenty-first century? *Annual Review of Earth and Planetary Sciences* **40**:89–111.

33   Bond, D.P.G. and S.E. Grasby. 2017. On the causes of mass extinctions.

34   Romm, J. 2015. *Climate Change*. Oxford University Press.

35   Intergovernmental Science-Policy Platform on Biodiversity and Ecosystem Services (IPBES). 2019. Media release. https://ipbes.net/news/Media-Release-Global-Assessment#1-Scale.

36   Boer, M.M., V. Resco de Dios and R.A. Bradstock. 2020. Unprecedented burn area of Australian mega forest fires. *Nature Climate Change* **10**:171–2.

37   https://www.theguardian.com/australia-news/2019/nov/24/world-heritage-queensland-rainforest-burned-for-10-days-and-almost-no-one-noticed (accessed 1 August 2020).

38   https://www.abc.net.au/radionational/programs/scienceshow/terania-creek---saved-in-1979,-ravaged-by-fire-in-2019/11988330 (accessed 1 August 2020).

**Chapter Four: A million years of rain**

1   Tillyard, R.J. 1908. On the genus *Petalura*, with description of a new species. *Proceedings of the Linnean Society of New South Wales* **32**:708–18.

2   Monteith, G. 2016. Have we got the heaviest dragonfly and can it fly in the wheel position? *Entomological Society of Queensland News Bulletin* **44**:106.

3   http://anic.ento.csiro.au/insectfamilies/biota_details.aspx?OrderID=24130&BiotaID=24215&PageID=families (accessed 1 June 2020).

4   Derhé, M.A., H. Murphy, G. Monteith and R. Menéndez. 2016. Measuring the success of reforestation for restoring biodiversity and ecosystem functioning. *Journal of Applied Ecology* **53**:1714–24.

5   Nel, A., P. Roques, P. Nel et al. 2013. The earliest known holometabolous insects. *Nature* **503**:257–61.

6   Misof, B., S. Liu, K. Meusemann et al. 2014. Phylogenomics resolves the timing and pattern of insect evolution. *Science* **346**:763–7.

7   Gunter, N.L., T.A. Weir, A. Slipinksi et al. 2016. If dung beetles (Scarabaeidae: Scarabaeinae) arose in association with dinosaurs, did they also suffer a mass co-extinction at the K-Pg boundary? *PLoS One* **11**:e0153570.

8    Carle, F.L., K.M. Kjer and M.L. May. 2015. A molecular phylogeny and classification of Anisoptera (Odonata). *Arthropod Systematics and Phylogeny* **73**:281–301.

9    Warren, A.A. 1980. *Parotosuchus* from the Early Triassic of Queensland and Western Australia. *Alcheringa: An Australasian Journal of Palaeontology* **4**:25–36.

10   Button, D.J., G.T. Lloyd, M.D. Ezcurra and R.J. Butler. 2017. Mass extinctions drove increased global faunal cosmopolitanism on the supercontinent Pangaea. *Nature Communications* **8**:733.

11   Pattemore, G.A. 2016. Megaflora of the Australian Triassic–Jurassic: A taxonomic revision. *Acta Palaeobotanica* **56**:121–82.

12   Retallack, G.J. 1977. Reconstructing Triassic vegetation of eastern Australasia: A new approach for the biostratigraphy of Gondwanaland. *Alcheringa: An Australasian Journal of Palaeontology* **1**:247–78.

13   Thulborn, T. 1998. Australia's earliest theropods: Footprint evidence in the Ipswich coal measures (Upper Triassic) of Queensland. *GAIA* **15**:301–11.

14   Brusatte, S.L., S.J. Nesbitt, R.B. Irmis et al. 2010. The origin and early radiation of dinosaurs. *Earth-Science Reviews* **101**:68–100.

15   Bernardi, M., P. Gianolla, F.M. Petti et al. 2018. Dinosaur diversification linked with the Carnian Pluvial Episode. *Nature Communications* **9**:1499.

16   Marshall, M. 2019. A million years of Triassic rain. *Nature* **576**:26–8.

17   Ogg, J.G. 2015. The mysterious mid-Carnian 'Wet Intermezzo' global event. *Journal of Earth Science* **26**:181–91.

18   Ogg, J.G. 2015. The mysterious mid-Carnian 'Wet Intermezzo' global event.

19   Benton, M.J., M. Bernardi and C. Kinsella. 2018. The Carnian Pluvial Episode and the origin of dinosaurs. *Journal of the Geological Society* **175**:1019–26.

20   Brusatte, S.L., S.J. Nesbitt, R.B. Irmis et al. 2010. The origin and early radiation of dinosaurs.

21   Bennett, S.C. 2020. Reassessment of the Triassic archosauriform *Scleromochlus taylori*: Neither runner nor biped, but hopper. *PeerJ* **8**:e8418.

22   https://en.wikipedia.org/wiki/Cryolophosaurus (accessed 1 June 2019).

23   Elliot, D.H., D. Larsen, C.M. Fanning et al. 2017. The Lower Jurassic Hanson Formation of the Transantarctic Mountains: Implications for the Antarctic sector of the Gondwana plate margin. *Geological Magazine* **154**:777–803.

24   McLoughlin, S., S.K. Martin and R. Beattie. 2015. The record of Australian Jurassic plant–arthropod interactions. *Gondwana Research* **27**:940–59.

25   Roelants, K., D.J. Gower, M. Wilkinson et al. 2007. Global patterns of diversification in the history of modern amphibians. *Proceedings of the National Academy of Sciences of the United States of America* **104**:887–92.

26   Romilio, A., S.W. Salisbury and A. Jannel. 2020. Footprints of large theropod dinosaurs in the Middle–Upper Jurassic (lower Callovian–lower Tithonian) Walloon Coal Measures of southern Queensland, Australia. *Historical Biology* **33**:1–12.

27   Kear, B.P. 2012. A revision of Australia's Jurassic plesiosaurs. *Palaeontology* **55**:1125–38.

28   Abdala, F. and A.M. Ribeiro. 2010. Distribution and diversity patterns of Triassic cynodonts (Therapsida, Cynodontia) in Gondwana. *Palaeogeography, Palaeoclimatology, Palaeoecology* **286**:202–17.

29   Luo, Z.-X. 2007. Transformation and diversification in early mammal evolution. *Nature* **450**:1011–19.

30   Gerkema, M.P., W.I.L. Davies, R.G. Foster et al. 2013. The nocturnal bottleneck and the evolution of activity patterns in mammals. *Proceedings of the Royal Society B: Biological Sciences.* **280**:20130508.

31 Close, R.A., M. Friedman, G.T. Lloyd and R.B.J. Benson. 2015. Evidence for a mid-Jurassic adaptive radiation in mammals. *Current Biology* 25:2137–42.

32 van Konijnenburg-van Cittert, J.H.A. 2008. The Jurassic fossil plant record of the UK area. *Proceedings of the Geologists' Association* 119:59–72.

33 Darwin, C. 2008. Letter to J.D. Hooker. 22 July 1879, in F. Darwin and A.C. Seward, editors. *More Letters of Charles Darwin*, Volume II. https://www. gutenberg.org/files/2740/2740-h/2740-h.htm#link2HCH0013.

34 Berry, C.M. 2019. Palaeobotany: The rise of the Earth's early forests. *Current Biology* 29:R792–R794.

35 Meyer-Berthaud, B., S.E. Scheckler and J. Wendt. 1999. Archaeopteris is the earliest known modern tree. *Nature* 398:700–1.

36 van Oosterzee, P. 1997. *Where Worlds Collide: The Wallace Line*. Reed.

37 Friedman, W.E. 2009. The meaning of Darwin's 'abominable mystery'. *American Journal of Botany* 96:5–21.

38 Silvestro, D., C.D. Bacon, W. Ding et al. 2021. Fossil data support a pre-Cretaceous origin of flowering plants. *Nature Ecology & Evolution* 5:449–57.

39 Cardinal, S. and B.N. Danforth. 2013. Bees diversified in the age of eudicots. *Proceedings of the Royal Society B: Biological Sciences* 280:20122686.

40 Cardinal, S. and B.N. Danforth. 2013. Bees diversified in the age of eudicots.

41 Armstrong, J.E. and A.K. Irvine. 1990. Functions of staminodia in the beetle-pollinated flowers of *Eupomatia laurina*. *Biotropica* 22:429–31.

42 The survey was conducted by Riegel Jensen.

43 Information was obtained by reference to the Australian Tropical Rainforest Plants key: https://apps.lucidcentral.org/rainforest/text/intro/index.html (accessed 1 June 2020).

44 http://whc.unesco.org/en/list/486/ (accessed 1 June 2020).

45 Friis, E.M., P.R. Crane and K.R. Pedersen. 2011. *Early Flowers and Angiosperm Evolution*. Cambridge University Press.

46 Stevens, P.F. 2022. Angiosperm Phylogeny Website. http://www.mobot.org/ MOBOT/research/APweb/welcome.html.

47 Silvestro, D., C.D. Bacon, W. Ding et al. 2021. Fossil data support a pre-Cretaceous origin of flowering plants.

48 Friis, E.M., P.R. Crane and K.R. Pedersen. 2011. *Early Flowers and Angiosperm Evolution*.

49 Feild, T.S., P.J. Franks and T.L. Sage. 2003. Ecophysiological shade adaptation in the basal angiosperm, *Austrobaileya scandens* (Austrobaileyaceae). *International Journal of Plant Sciences* 164:313–24.

50 Jacobs, M., M. López-García, O.-P. Phrathep et al. 2016. Photonic multilayer structure of *Begonia* chloroplasts enhances photosynthetic efficiency. *Nature Plants* 2:16162.

51 Sauquet, H., M. von Balthazar, S. Magallón et al. 2017. The ancestral flower of angiosperms and its early diversification. *Nature Communications* 8:16047.

52 Feild, T.S., P.J. Franks and T.L. Sage. 2003. Ecophysiological shade adaptation in the basal angiosperm, *Austrobaileya scandens* (Austrobaileyaceae).

53 Barrett, P.M. and K.J. Willis. 2001. Did dinosaurs invent flowers? Dinosaur–angiosperm coevolution revisited. *Biological reviews of the Cambridge Philosophical Society* 76:411–47.

54 Eriksson, O., E.M. Friis and P. Löfgren. 2000. Seed size, fruit size, and dispersal systems in angiosperms from the Early Cretaceous to the Late Tertiary. *The American Naturalist* 156:47–58.

55  Morrissey, P. 2015. Bill Neidjie's *Story About Feeling*: Notes on its themes and philosophy. *Journal of the Association for the Study of Australian Literature* **15**:1.

56  The poem is published in Neidjie, B. 1989. *Story About Feeling*. K. Taylor, editor. Magabala Books.

57  https://evolution.berkeley.edu/it-takes-teamwork-how-endosymbiosis-changed-life-on-earth/evidence-for-endosymbiosis/ (accessed 1 June 2020).

58  Calvo, P., V.P. Sahi and A. Trewavas. 2017. Are plants sentient? *Plant, Cell & Environment* **40**:2858–69.

59  Trewavas, A. 2005. Plant intelligence. *Naturwissenschaften* **92**:401–13.

60  Wet Tropics Management Authority. 2014. State of Wet Tropics Report 2013/14: Ancient, threatened and endemic plants of the Wet Tropics World Heritage Area.

61  http://www.mobot.org/MOBOT/research/APweb/ (accessed 1 June 2020); https://www.newscientist.com/article/dn17453-timeline-the-evolution-of-life/ (accessed 1 June 2020).

62  McLoughlin, S. 2001. The breakup history of Gondwana and its impact on pre-Cenozoic floristic provincialism. *Australian Journal of Botany* **49**:271–300.

63  McLoughlin, S. and B.P. Kear. 2010. The Australasian Cretaceous scene. *Alcheringa: An Australasian Journal of Palaeontology* **34**:197–203.

64  Rich, T.H., P. Vickers-Rich, T.F. Flannery et al. 2009. An Australian multituberculate and its palaeobiogeographic implications. *Acta Palaeontologica Polonica* **54**:1–6.

65  https://australian.museum/learn/australia-over-time/evolving-landscape/the-cretaceous-period/ (accessed 1 June 2020).

66  McLoughlin, S. 2001. The breakup history of Gondwana and its impact on pre-Cenozoic floristic provincialism.

67  Close, R.A., P. Vickers-Rich, P. Trusler et al. 2009. Earliest Gondwanan bird from the Cretaceous of southeastern Australia. *Journal of Vertebrate Paleontology* **29**:616–19.

68  http://www.mobot.org/MOBOT/research/APweb/ (accessed 1 June 2020).

69  Dettmann, M.E. 1994. Cretaceous vegetation: The microfossil record. pp. 143–70 in R.S. Hill, editor. *History of the Australian Vegetation*. University of Adelaide Press.

70  Dettmann, M.E., R.E. Molnar, J.G. Douglas et al. 1992. Australian Cretaceous terrestrial faunas and floras: Biostratigraphic and biogeographic implications. *Cretaceous Research* **13**:207–62.

## Chapter Five: Zombie busters

1  Webb, L.J. 1966. The rape of the forests. pp. 178–228 in A.J. Marshall, editor. *The Great Extermination*. Heinemann.

2  Frawley, K.J. 1983. Forest and land management in north-east Queensland: 1859–1960. PhD thesis. ANU.

3  Winter, J.W., F.C. Bell, L.I. Pahl and R.G. Atherton. 1987. Rainforest clearing in northeastern Australia. *Proceedings of the Royal Society of Queensland* **98**:41–57.

4  Conrad, J. 2002 (original 1899). *Heart of Darkness and Other Tales*. Oxford University Press.

5  Webb, L.J. 1966. The rape of the forests.

6  http://www.in2013dollars.com/1870-GBP-in-2016 (accessed 1 June 2021).

7  http://mpegmedia.abc.net.au/rn/podcast/2011/07/hht_20110724.mp3 (accessed 31 July 2022).

8  A super foot is 1/12 of a cubic foot, so this is 1.25 million cubic feet. There are 35.3 cubic feet in a cubic metre, so 1.25 million divided by 35.3 is 35,411, which is the approximate number of trees if a tree at 40 centimetres diameter at breast height (dbh) is equal to 1 cubic metre.

9 Frawley, K.J. 1983. Forest and land management in north-east Queensland: 1859–1960. PhD thesis. ANU.

10 Boyd, A.J. 1886. Queensland: An introductory essay. *Essays prepared for the Colonial and Indian Exhibition, London.*

11 Field, D.J., A. Bercovici, J.S. Berv et al. 2018. Early evolution of modern birds structured by global forest collapse at the end-Cretaceous mass extinction. *Current Biology* **28**:1825–31.e2.

12 Brusatte, S.L., R.J. Butler, P.M. Barrett et al. 2015. The extinction of the dinosaurs. *Biological Reviews* **90**:628–42.

13 https://www.livescience.com/64426-dinosaur-killing-asteroid-caused-giant-tsunami.html (accessed 1 June 2020).

14 Brugger, J., G. Feulner and S. Petri. 2017. Baby, it's cold outside: Climate model simulations of the effects of the asteroid impact at the end of the Cretaceous. *Geophysical Research Letters* **44**:419–27.

15 Longrich, N.R., T. Tokaryk and D.J. Field. 2011. Mass extinction of birds at the Cretaceous–Paleogene (K-Pg) boundary. *Proceedings of the National Academy of Sciences* **108**:15253–7.

16 Longrich, N.R., B.-A.S. Bhullar and J.A. Gauthier. 2012. Mass extinction of lizards and snakes at the Cretaceous–Paleogene boundary. *Proceedings of the National Academy of Sciences* **109**:21396–401.

17 Chiarenza, A.A., A. Farnsworth, P.D. Mannion et al. 2020. Asteroid impact, not volcanism, caused the end-Cretaceous dinosaur extinction. *Proceedings of the National Academy of Sciences* **117**:17084–93.

18 Longrich, N.R., B.-A.S. Bhullar and J.A. Gauthier. 2012. Mass extinction of lizards and snakes at the Cretaceous–Paleogene boundary.

19 Field, D.J., A. Bercovici, J.S. Berv et al. 2018. Early evolution of modern birds structured by global forest collapse at the end-Cretaceous mass extinction.

20 Ceballos, G., P.R. Ehrlich and P.H. Raven. 2020. Vertebrates on the brink as indicators of biological annihilation and the sixth mass extinction. *Proceedings of the National Academy of Sciences* **117**:13596–602.

21 Mjöberg, E. 2015. *Amongst Stone Age People in the Queensland Wilderness.* Hesperian Press. p. 89.

22 Mjöberg, E. 2015. *Amongst Stone Age People in the Queensland Wilderness.* p. 89.

23 Laurance, W.F., S.G. Laurance and D.W. Hilbert. 2008. Long-term dynamics of a fragmented rainforest mammal assemblage. *Conservation Biology* **22**:1154–64.

24 These figures extrapolated from Laurance, W.F., S.G. Laurance and D.W. Hilbert. 2008. Long-term dynamics of a fragmented rainforest mammal assemblage.

25 McCallum, M.L. 2015. Vertebrate biodiversity losses point to a sixth mass extinction. *Biodiversity and Conservation* **24**:2497–519.

26 McCallum, M.L. 2015. Vertebrate biodiversity losses point to a sixth mass extinction.

27 Ceballos, G., P.R. Ehrlich and P.H. Raven. 2020. Vertebrates on the brink as indicators of biological annihilation and the sixth mass extinction.

28 Díaz, S., J. Settele, E.S. Brondízio et al. 2019. Pervasive human-driven decline of life on Earth points to the need for transformative change. *Science* **366**:eaax3100.

29 Bologna, M. and G. Aquino. 2020. Deforestation and world population sustainability: A quantitative analysis. *Scientific Reports* **10**:7631.

30 Bologna, M. and G. Aquino. 2020. Deforestation and world population sustainability.

31 Ceballos, G., P.R. Ehrlich and P.H. Raven. 2020. Vertebrates on the brink as indicators of biological annihilation and the sixth mass extinction.

## Chapter Six: The walk

1   Kolomyjec, S.H., T.R. Grant, C.N. Johnson and D. Blair. 2013. Regional population structuring and conservation units in the platypus (*Ornithorhynchus anatinus*). *Australian Journal of Zoology* **61**:378–85.

2   Schodde, R. and S. Tidemann, editors. 2010. *Reader's Digest Complete Book of Australian Birds*. Reader's Digest.

3   Andersen, M.J., Á.S. Nyári, I. Mason et al. 2014. Molecular systematics of the world's most polytypic bird: The *Pachycephala pectoralis/melanura* (Aves: Pachycephalidae) species complex. *Zoological Journal of the Linnean Society* **170**:566–88.

4   Cornetti, L., L.M. Valente, L.T. Dunning et al. 2015. The genome of the 'great speciator' provides insights into bird diversification. *Genome Biology and Evolution* 7:2680–91.

5   Cornetti, L., L.M. Valente, L.T. Dunning et al. 2015. The genome of the 'great speciator' provides insights into bird diversification.

6   https://www.wettropics.gov.au/birds (accessed 12 October 2020).

7   Hilbert, D.W., M. Bradford, T. Parker and D.A. Westcott. 2004. Golden bowerbird (*Prionodura newtonia*) habitat in past, present and future climates: Predicted extinction of a vertebrate in tropical highlands due to global warming. *Biological Conservation* **116**:367–77.

8   Norman, J.A., L. Christidis and R. Schodde. 2018. Ecological and evolutionary diversification in the Australo-Papuan scrubwrens (*Sericornis*) and mouse-warblers (*Crateroscelis*), with a revision of the subfamily Sericornithinae (Aves: Passeriformes: Acanthizidae). *Organisms Diversity & Evolution* **18**:241–59.

9   Koetz, A.H., D.A. Westcott and B.C. Congdon. 2007. Geographical variation in song frequency and structure: The effects of vicariant isolation, habitat type and body size. *Animal Behaviour* **74**:1573–83.

10  Aggerbeck, M., J. Fjeldså, L. Christidis et al. 2014. Resolving deep lineage divergences in core corvoid passerine birds supports a proto-Papuan island origin. *Molecular Phylogenetics and Evolution* **70**:272–85.

11  Brusatte, S.L. 2016. Evolution: How some birds survived when all other dinosaurs died. *Current Biology* **26**:R415–R417.

12  Brusatte, S.L., G.T. Lloyd, S.C. Wang and M.A. Norell. 2014. Gradual assembly of avian body plan culminated in rapid rates of evolution across the dinosaur–bird transition. *Current Biology* **24**:2386–92.

13  Larson, D.W., C.M. Brown and D.C. Evans. 2016. Dental disparity and ecological stability in bird-like dinosaurs prior to the end-Cretaceous mass extinction. *Current Biology* **26**:1325–33.

14  https://whc.unesco.org/en/list/402/ (accessed 12 October 2020).

15  Maderspacher, F. 2017. Evolution: Flight of the ratites. *Current Biology* **27**:R110–R113.

16  https://www.nationalgeographic.com/news/2014/5/140513-flightless-birds-ostriches-moas-evolution-science/ (accessed 12 October 2020).

17  Yonezawa, T., T. Segawa, H. Mori et al. 2017. Phylogenomics and morphology of extinct paleognaths reveal the origin and evolution of the ratites. *Current Biology* **27**:68–77.

18  Phillips, M.J., G.C. Gibb, E.A. Crimp and D. Penny. 2010. Tinamous and moa flock together: Mitochondrial genome sequence analysis reveals independent losses of flight among ratites. *Systematic Biology* **59**:90–107.

19   Wright, N.A., D.W. Steadman and C.C. Witt. 2016. Predictable evolution toward flightlessness in volant island birds. *Proceedings of the National Academy of Sciences* **113**:4765–70.

20   Eastick, D.L., G.J. Tattersall, S.J. Watson et al. 2019. Cassowary casques act as thermal windows. *Scientific Reports* **9**:1966.

21   Bradford, M.G., A.J. Dennis and D.A. Westcott. 2008. Diet and dietary preferences of the southern cassowary (*Casuarius casuarius*) in North Queensland, Australia. *Biotropica* **40**:338–43.

22   Latch, P. 2008. *Recovery plan for Mabi Forest. Report to Department of the Environment, Water, Heritage and the Arts Canberra.* Environment Protection Agency, Brisbane.

23   Bradford, M.G. and D.A. Westcott. 2010. Consequences of southern cassowary (*Casuarius casuarius*, L.) gut passage and deposition pattern on the germination of rainforest seeds. *Austral Ecology* **35**:325–33.

24   Latch, P. 2008. *Recovery plan for Mabi Forest.*

## Chapter Seven: Mabi's world

1    Rose, D.B. 2013. Death and grief in a world of kin. pp. 137–47 in G. Harvey, editor. *The Handbook of Contemporary Animism.* Routledge.

2    Elkin, A.P. 1961. The Yabuduruwa. *Oceania* **31**:166–209.

3    Roberts, P., A. Buhrich, V. Caetano-Andrade et al. 2021. Reimagining the relationship between Gondwanan forests and Aboriginal land management in Australia's 'Wet Tropics', *iScience* **24**:102190.

4    Rose, D.B. 2000. To Dance with time: A Victoria River Aboriginal study. *The Australian Journal of Anthropology* **11**:287–96.

5    Pannell, S. 2006. *Yamani Country: A spatial history of the Atherton Tableland, North Queensland.* Rainforest CRC. p. 93.

6    Pannell, S. 2006. *Yamani Country.* p. 93.

7    Dixon, R.M.W. 2017. *Dyirbal Texts: 78 legends, stories, autobiographies, conversations, and remedies in Jirrbal, Girramay, Mamu, and Gulngay.* Language and Culture Research Centre, James Cook University.

8    The scene comes from Govor, E. 2000. *My Dark Brother: The story of the Illins, a Russian–Aboriginal Family.* UNSW Press.

9    This version of the story is told by Warren Canendo in Pannell, S. 2006. *Yamani Country.*

10   Confirmation made by Professor David Gillieson.

11   Bottoms, T. 2013. *Conspiracy of Silence: Queensland's frontier killing times.* Allen & Unwin. p. 150.

12   The version I used came from Dixon, R.M.W. 2017. *Dyirbal Texts.*

13   Whitehead, P.W., P.J. Stephenson, I. McDougall et al. 2007. Temporal development of the Atherton Basalt Province, north Queensland. *Australian Journal of Earth Sciences* **54**:691–709.

14   Whitehead, P.W., P.J. Stephenson, I. McDougall et al. 2007. Temporal development of the Atherton Basalt Province, north Queensland.

15   Neal, C.A., S.R. Brantley, L. Antolik et al. 2019. The 2018 rift eruption and summit collapse of Kīlauea Volcano. *Science* **363**:367–74.

16   Pannell, S. 2006. *Yamani Country.*

17   Müller, R.D., N. Flament, K.J. Matthews et al. 2016. Formation of Australian continental margin highlands driven by plate–mantle interaction. *Earth and Planetary Science Letters* **441**:60–70.

18 https://www.wettropics.gov.au/site/user-assets/docs/63WetTropicsGeology.pdf (accessed 16 April 2020).

19 Cohen, B.E., D.F. Mark, S.J. Fallon and P.J. Stephenson. 2017. Holocene–Neogene volcanism in northeastern Australia: Chronology and eruption history. *Quaternary Geochronology* **39**:79–91.

20 Baldwin, S.L., P.G. Fitzgerald and L.E. Webb. 2012. Tectonics of the New Guinea region. *Annual Review of Earth and Planetary Sciences* **40**:495–520.

21 Cited in Pannell, S. 2006. *Yamani Country.*

22 Interview with Ernie Raymont, 24 November 2020.

23 Black, K.H., M. Archer, S.J. Hand and H. Godthelp. 2012. The rise of Australian marsupials: A synopsis of biostratigraphic, phylogenetic, palaeoecologic and palaeo-biogeographic understanding. pp. 983–1078 in J.A. Talent, editor. *Earth and Life.* Springer.

24 Luo, Z.-X., Q. Ji, J.R. Wible and C.-X. Yuan. 2003. An early cretaceous tribosphenic mammal and metatherian evolution. *Science* 302:1934–40.

25 Gelfo, J.N., T. Mörs, M. Lorente et al. 2015. The oldest mammals from Antarctica, early Eocene of the La Meseta Formation, Seymour Island. *Palaeontology* **58**:101–10.

26 Geiser, F. and G. Körtner. 2010. Hibernation and daily torpor in Australian mammals. *Australian Zoologist* **35**:204–15.

27 Beck, R.M.D. 2008. Form, function, phylogeny and biogeography of enigmatic Australian metatherians. PhD thesis. University of New South Wales.

28 Godthelp, H., M. Archer, R. Cifelli et al. 1992. Earliest known Australian Tertiary mammal fauna. *Nature* **356**:514–16.

29 Beck, R.M.D. 2008. Form, function, phylogeny and biogeography of enigmatic Australian metatherians. PhD thesis. University of New South Wales.

30 Boles, W.E. 1997. Fossil songbirds (Passeriformes) from the early Eocene of Australia. *Emu: Austral Ornithology* **97**:43–50.

31 Sibley, C.G., J.E. Ahlquist and B.L. Monroe Jr. 1988. A classification of the living birds of the world, based on DNA–DNA hybridization studies. *The Auk* **105**: 409–23.

32 Claramunt, S. and J. Cracraft. 2015. A new time tree reveals Earth history's imprint on the evolution of modern birds. *Science Advances* **1**:e1501005.

33 Morse, P.E., S.G.B. Chester, D.M. Boyer et al. 2019. New fossils, systematics, and biogeography of the oldest known crown primate *Teilhardina* from the earliest Eocene of Asia, Europe, and North America. *Journal of Human Evolution* **128**:103–31.

34 https://en.wikipedia.org/wiki/Omomyidae (accessed 16 April 2020).

35 McInerney, F.A. and S.L. Wing. 2011. The Paleocene–Eocene Thermal Maximum: A perturbation of carbon cycle, climate, and biosphere with implications for the future. *Annual Review of Earth and Planetary Sciences* **39**:489–516.

36 Kooyman, R.M., R.J. Morley, D.M. Crayn et al. 2019. Origins and assembly of Malesian rainforests. *Annual Review of Ecology, Evolution, and Systematics* **50**:119–43.

37 Greenwood, D.R. and D.C. Christophel. 2005. The origins and Tertiary history of Australian 'tropical' rainforests. pp. 336–73 in E. Bermingham, C. Dick and C. Moritz, editors. *Tropical Rainforests: Past, present, and future.* University of Chicago Press.

38 Ladiges, P.Y., F. Udovicic and G. Nelson. 2003. Australian biogeographical connections and the phylogeny of large genera in the plant family Myrtaceae. *Journal of Biogeography* **30**:989–98.

39 Greenwood, D.R. and D.C. Christophel. 2005. The origins and Tertiary history of Australian 'tropical' rainforests.

40    Black, K.H., M. Archer, S.J. Hand and H. Godthelp. 2012. The rise of Australian marsupials.

41    Beck, R.M.D., N.M. Warburton, M. Archer et al. 2016. Going underground: Postcranial morphology of the early Miocene marsupial mole *Naraboryctes philcreaseri* and the evolution of fossoriality in notoryctemorphians. *Memoirs of Museum Victoria* **74**:151–71.

42    Eriksson, O. 2008. Evolution of seed size and biotic seed dispersal in angiosperms: Paleoecological and neoecological evidence. *International Journal of Plant Sciences* **169**:863–70.

43    Tyler, M.J. and H. Godthelp. 1993. A new species of *Lechriodus* Boulenger (Anura: Leptodactylidae) from the early Eocene of Queensland. *Transactions of the Royal Society of South Australia* **117**:187–9.

44    Worthy, T.H., F.J. Degrange, W.D. Handley and M.S.Y. Lee. 2017. The evolution of giant flightless birds and novel phylogenetic relationships for extinct fowl (Aves, Galloanseres). *Royal Society Open Science* **4**:170975.

45    https://www.wikiwand.com/en/Gastornis (accessed 16 April 2020).

46    Archer, M., S. Hand and H. Godthelp. 1991. *Riversleigh: The story of animals in ancient rainforests of inland Australia.* Reed.

47    Clarkson, C., Z. Jacobs, B. Marwick et al. 2017. Human occupation of northern Australia by 65,000 years ago. *Nature* **547**:306–10.

48    https://en.wikipedia.org/wiki/Dromornis (accessed 16 April 2020).

49    Black, K.H., M. Archer, S.J. Hand and H. Godthelp. 2012. The rise of Australian marsupials.

50    Archer, M., S. Hand and H. Godthelp. 1991. *Riversleigh*.

51    Black, K.H., M. Archer, S.J. Hand and H. Godthelp. 2012. The rise of Australian marsupials.

52    Archer, M., S. Hand and H. Godthelp. 1991. *Riversleigh*.

53    Moyle, R.G., C.H. Oliveros, M.J. Andersen et al. 2016. Tectonic collision and uplift of Wallacea triggered the global songbird radiation. *Nature Communications* **7**:12709.

54    Holbourn, A., W. Kuhnt, K.G.D. Kochhann et al. 2015. Global perturbation of the carbon cycle at the onset of the Miocene Climatic Optimum. *Geology* **43**:123–6.

55    https://en.wikipedia.org/wiki/Columbia_River_Basalt_Group (accessed 1 June 2020).

56    Black, K.H., M. Archer, S.J. Hand and H. Godthelp. 2012. The rise of Australian marsupials.

57    Holbourn, A.F., W. Kuhnt, S.C. Clemens et al. 2018. Late Miocene climate cooling and intensification of southeast Asian winter monsoon. *Nature Communications* **9**:1584.

58    Lyle, M., J. Barron, T.J. Bralower et al. 2008. Pacific Ocean and Cenozoic evolution of climate. *Reviews of Geophysics* **46**:RG2002.

59    Crayn, D.M., C. Costion and M.G. Harrington. 2015. The Sahul–Sunda floristic exchange: Dated molecular phylogenies document Cenozoic intercontinental dispersal dynamics. *Journal of Biogeography* **42**:11–24.

60    https://www.nature.com/scitable/knowledge/library/hominoid-origins-135874580/ (accessed 1 September 2020).

61    https://www.scientificamerican.com/article/planet-of-the-apes-2006-06/ (accessed 1 September 2020 ).

62    Hall, R. 2013. The palaeogeography of Sundaland and Wallacea since the Late Jurassic. *Journal of Limnology* **72**:1–17.

63  Kuhl, H., C. Frankl-Vilches, A. Bakker et al. 2021. An unbiased molecular approach using 3′-UTRs resolves the avian family-level tree of life. *Molecular Biology and Evolution* **38**:108–27.

64  Martin, H.A. 2006. Cenozoic climatic change and the development of the arid vegetation in Australia. *Journal of Arid Environments* **66**:533–63.

65  Arnaiz-Villena, A., V. Ruiz-del-Valle, P. Gomez-Prieto et al. 2009. Estrildinae finches (Aves, Passeriformes) from Africa, South Asia and Australia: A molecular phylogeographic study. *The Open Ornithology Journal* **2**:29–36.

66  Olsson, U. and P. Alström. 2020. A comprehensive phylogeny and taxonomic evaluation of the waxbills (Aves: Estrildidae). *Molecular Phylogenetics and Evolution* **146**:106757.

67  Andersen, M.J., J.M. McCullough, W.M. Mauck III et al. 2018. A phylogeny of kingfishers reveals an Indomalayan origin and elevated rates of diversification on oceanic islands. *Journal of Biogeography* **45**:269–81.

68  Hocknull, S.A. 2009. Late Cainozoic rainforest vertebrates from Australopapua: Evolution, biogeography and extinction. PhD thesis. University of New South Wales.

69  Yap, J.-Y.S., M. Rossetto, C. Costion et al. 2018. Filters of floristic exchange: How traits and climate shape the rain forest invasion of Sahul from Sunda. *Journal of Biogeography* **45**:838–47.

70  Yap, J.-Y.S., M. Rossetto, C. Costion et al. 2018. Filters of floristic exchange.

71  Holbourn, A.E., W. Kuhnt, S.C. Clemens et al. 2018. Late Miocene climate cooling and intensification of southeast Asian winter monsoon.

72  Lawver, L.A. and L.M. Gahagan. 2003. Evolution of Cenozoic seaways in the circum-Antarctic region. *Palaeogeography, Palaeoclimatology, Palaeoecology* **198**:11–37.

73  Meredith, R.W., M. Westerman and M.S. Springer. 2009. A phylogeny and timescale for the living genera of kangaroos and kin (Macropodiformes: Marsupialia) based on nuclear DNA sequences. *Australian Journal of Zoology* **56**:395–410.

74  Eldridge, M.D.B., S. Potter, K.M. Helgen et al. 2018. Phylogenetic analysis of the tree-kangaroos (*Dendrolagus*) reveals multiple divergent lineages within New Guinea. *Molecular Phylogenetics and Evolution* **127**:589–99.

75  https://www.tree-kangaroo.net/tree-kangaroos-australia-and-png/tree-kangaroo-overview (accessed 12 January 2021).

76  Karas, C., D. Nürnberg, A. Bahr et al. 2017. Pliocene oceanic seaways and global climate. *Scientific Reports* **7**:39842.

77  Burke, K.D., J.W. Williams, M.A. Chandler et al. 2018. Pliocene and Eocene provide best analogs for near-future climates. *Proceedings of the National Academy of Sciences* **115**:13288–93.

78  This description of Thiaki is based on Hocknull, S. 2005. Ecological succession during the late Cainozoic of central eastern Queensland: Extinction of a diverse rainforest community. *Memoirs of the Queensland Museum* **51**:39–122.

79  Prideaux, G.J. and N.M. Warburton. 2010. An osteology–based appraisal of the phylogeny and evolution of kangaroos and wallabies (Macropodidae: Marsupialia). *Zoological Journal of the Linnean Society* **159**:954–87.

80  Eldridge, M.D.B., S. Potter, K.M. Helgen et al. 2018. Phylogenetic analysis of the tree-kangaroos (*Dendrolagus*) reveals multiple divergent lineages within New Guinea.

81  White, T.D., C.O. Lovejoy, B. Asfaw et al. 2015. Neither chimpanzee nor human, *Ardipithecus* reveals the surprising ancestry of both. *Proceedings of the National Academy of Sciences* **112**:4877–84.

82    Karas, C., D. Nürnberg, A. Bahr et al. 2017. Pliocene oceanic seaways and global climate.

83    Köhler, P. and R.S.W. van de Wal. 2020. Interglacials of the Quaternary defined by northern hemispheric land ice distribution outside of Greenland. *Nature Communications* **11**:5124.

84    Williams, A.N., S. Ulm, T. Sapienza et al. 2018. Sea-level change and demography during the last glacial termination and early Holocene across the Australian continent. *Quaternary Science Reviews* **182**:144–54.

## Chapter Eight: Nothing is nothing

1     Van Arsdale, A.P. 2013. *Homo erectus*: A bigger, smarter, faster hominin lineage. *Nature Education Knowledge* **4**:2.

2     Dennell, R. and W. Roebroeks. 2005. An Asian perspective on early human dispersal from Africa. *Nature* **438**:1099–104.

3     Wurster, C.M., H. Rifai, B. Zhou et al. 2019. Savanna in equatorial Borneo during the late Pleistocene. *Scientific Reports* **9**:6392.

4     Morwood, M. and P. van Oosterzee. 2007. *The Discovery of the Hobbit: The scientific breakthrough that changed the face of human history*. Random House.

5     Spratt, R.M. and L.E. Lisiecki. 2016. A late Pleistocene sea level stack. *Climate of the Past* **12**:1079–92.

6     Bergström, A., C. Stringer, M. Hajdinjak et al. 2021. Origins of modern human ancestry. *Nature* **590**:229–37.

7     Rogers, A.R., N.S. Harris and A.A. Achenbach. 2020. Neanderthal–Denisovan ancestors interbred with a distantly related hominin. *Science Advances* **6**:eaay5483.

8     Krause, J., Q. Fu, J.M. Good et al. 2010. The complete mitochondrial DNA genome of an unknown hominin from southern Siberia. *Nature* **464**:894–7.

9     https://www.nature.com/articles/d41586-018-06004-0#ref-CR1 (accessed 10 February 2021).

10    Kaifu, Y. 2017. Archaic hominin populations in Asia before the arrival of modern humans. *Current Anthropology* **58**:S418–S433.

11    Kissel, M. and A. Fuentes. 2021. The ripples of modernity: How we can extend paleoanthropology with the extended evolutionary synthesis. *Evolutionary Anthropology* **30**:84–98.

12    Hocknull, S. 2005. Ecological succession during the late Cainozoic of central eastern Queensland: Extinction of a diverse rainforest community. *Memoirs of the Queensland Museum* **51**:39–122.

13    Hocknull, S.A., J.-x. Zhao, Y.-x. Feng and G.E. Webb. 2007. Responses of Quaternary rainforest vertebrates to climate change in Australia. *Earth and Planetary Science Letters* **264**:317–31.

14    Bergström, A., C. Stringer, M. Hajdinjak et al. 2021. Origins of modern human ancestry. *Nature* **590**:229–37.

15    Westaway, K.E., J. Louys, R.D. Awe et al. 2017. An early modern human presence in Sumatra 73,000–63,000 years ago. *Nature* **548**:322–5.

16    Rizal, Y., K.E. Westaway, Y. Zaim et al. 2020. Last appearance of *Homo erectus* at Ngandong, Java, 117,000–108,000 years ago. *Nature* **577**:381–5.

17    Louys, J. and P. Roberts. 2020. Environmental drivers of megafauna and hominin extinction in Southeast Asia. *Nature* **586**:402–6.

18    Jacobs, G.S., G. Hudjashov, L. Saag et al. 2019. Multiple deeply divergent Denisovan ancestries in Papuans. *Cell* **177**:1010–1021.e32.

19  Westaway, K.E., J. Louys, R.D. Awe et al. 2017. An early modern human presence in Sumatra 73,000–63,000 years ago.

20  Chivas, A.R., A. García, S. van der Kaars et al. 2001. Sea-level and environmental changes since the last interglacial in the Gulf of Carpentaria, Australia: An overview. *Quaternary International* **83–85**:19–46.

21  Bird, M.I., R.J. Beaman, S.A. Condie et al. 2018. Palaeogeography and voyage modeling indicates early human colonization of Australia was likely from Timor-Roti. *Quaternary Science Reviews* **191**:431–9.

22  Clarkson, C., Z. Jacobs, B. Marwick et al. 2017. Human occupation of northern Australia by 65,000 years ago. *Nature* **547**:306–10.

23  Bird, M.I., S.A. Condie, S. O'Connor et al. 2019. Early human settlement of Sahul was not an accident. *Scientific Reports* **9**:8220.

24  Bradshaw, C.J.A., S. Ulm, A.N. Williams et al. 2019. Minimum founding populations for the first peopling of Sahul. *Nature Ecology & Evolution* **3**:1057–63.

25  Pedro, N., N. Brucato, V. Fernandes et al. 2020. Papuan mitochondrial genomes and the settlement of Sahul. *Journal of Human Genetics* **65**:875–87.

26  Denham, T., R. Fullagar and L. Head. 2009. Plant exploitation on Sahul: From colonisation to the emergence of regional specialisation during the Holocene. *Quaternary International* **202**:29–40.

27  Bradshaw, C.J.A., K. Norman, S. Ulm et al. 2021. Stochastic models support rapid peopling of Late Pleistocene Sahul. *Nature Communications* **12**:2440.

28  Florin, S.A., P. Roberts, B. Marwick et al. 2021. Pandanus nutshell generates a palaeoprecipitation record for human occupation at Madjedbebe, northern Australia. *Nature Ecology & Evolution* **5**:295–303.

29  Cohen, T.J., G.C. Nanson, J.D. Jansen et al. 2011. Continental aridification and the vanishing of Australia's megalakes. *Geology* **39**:167–70.

30  Tobler, R., A. Rohrlach, J. Soubrier et al. 2017. Aboriginal mitogenomes reveal 50,000 years of regionalism in Australia. *Nature* **544**:180–4.

31  Crabtree, S.A., D.A. White, C.J.A. Bradshaw et al. 2021. Landscape rules predict optimal superhighways for the first peopling of Sahul. *Nature Human Behaviour* **5**:1303–13.

32  Hocknull, S.A., R. Lewis, L.J. Arnold et al. 2020. Extinction of eastern Sahul megafauna coincides with sustained environmental deterioration. *Nature Communications* **11**:2250.

33  Moss, P.T., G.B. Dunbar, Z. Thomas et al. 2017. A 60,000–year record of environmental change for the Wet Tropics of north–eastern Australia based on the ODP 820 marine core. *Journal of Quaternary Science* **32**:704–16.

34  Kemp, C.W., J. Tibby, L.J. Arnold and C. Barr. 2019. Australian hydroclimate during Marine Isotope Stage 3: A synthesis and review. *Quaternary Science Reviews* **204**:94–104.

35  Hocknull, S.A., R. Lewis, L.J. Arnold et al. 2020. Extinction of eastern Sahul megafauna coincides with sustained environmental deterioration.

36  Clarkson, C., Z. Jacobs, B. Marwick et al. 2017. Human occupation of northern Australia by 65,000 years ago.

37  Morrissey, P. 2015. Bill Neidjie's *Story About Feeling*: Notes on its themes and philosophy. *Journal of the Association for the Study of Australian Literature* **15**:1.

38  Mowaljarlai, D. and J. Malnic. 1993. *Yorro Yorro: Original creation and renewal of nature*. Magabala Books.

39  Williams, A.N., S. Ulm, A.R. Cook et al. 2013. Human refugia in Australia during the Last Glacial Maximum and Terminal Pleistocene: A geospatial analysis of the

25–12 ka Australian archaeological record. *Journal of Archaeological Science* **40**: 4612–25.

40  Finch, D., A. Gleadow, J. Hergt et al. 2020. 12,000-year-old Aboriginal rock art from the Kimberley region, Western Australia. *Science Advances* **6**:eaay3922.

41  Akerman, K. 2009. Interaction between humans and megafauna depicted in Australian rock art? *Antiquity* **83**:1–4.

42  https://australian.museum/learn/animals/mammals/thylacoleo-carnifex/ (accessed 10 February 2021).

43  Akerman, K. 2009. Interaction between humans and megafauna depicted in Australian rock art?

44  Williams, A.N., S. Ulm, T. Sapienza et al. 2018. Sea-level change and demography during the last glacial termination and early Holocene across the Australian continent. *Quaternary Science Reviews* **182**:144–54.

45  Reeves, J.M., H.C. Bostock, L.K. Ayliffe et al. 2013. Palaeoenvironmental change in tropical Australasia over the last 30,000 years: A synthesis by the OZ-INTIMATE group. *Quaternary Science Reviews* **74**:97–114.

46  Finch, D., A. Gleadow, J. Hergt et al. 2020. 12,000-year-old Aboriginal rock art from the Kimberley region, Western Australia.

47  Mowaljarlai, D. and J. Malnic. 1993. *Yorro Yorro*.

48  Moss, P.T., G.B. Dunbar, Z. Thomas et al. 2017. A 60000–year record of environmental change for the Wet Tropics of north–eastern Australia based on the ODP 820 marine core. *Journal of Quaternary Science* **32**:704–16.

49  Hoover, D.L., A.K. Knapp and M.D. Smith. 2014. Resistance and resilience of a grassland ecosystem to climate extremes. *Ecology* **95**:2646–56.

50  The story of Ngunya comes from two sources that tell quite different versions: Nunn, P.D. and N.J. Reid. 2016. Aboriginal memories of inundation of the Australian coast dating from more than 7000 years ago. *Australian Geographer* **47**:11–47; Bottoms, T. 2015. *Cairns City of the South Pacific: A history 1770–1995*. Bunu Bunu Press.

51  Dixon, R.M.W. 2017. *Dyirbal Texts: 78 legends, stories, autobiographies, conversations, and remedies in Jirrbal, Girramay, Mamu, and Gulngay*. Language and Culture Research Centre, James Cook University.

52  Williams, A.N., S. Ulm, T. Sapienza et al. 2018. Sea-level change and demography during the last glacial termination and early Holocene across the Australian continent.

53  Williams, A.N., S. Ulm, A.R. Cook et al. 2013. Human refugia in Australia during the Last Glacial Maximum and Terminal Pleistocene.

54  Florin, S.A., P. Roberts, B. Marwick et al. 2021. Pandanus nutshell generates a palaeoprecipitation record for human occupation at Madjedbebe, northern Australia.

55  Florin, S.A., A.S. Fairbairn, M. Nango et al. 2020. The first Australian plant foods at Madjedbebe, 65,000–53,000 years ago. *Nature Communications* **11**:924.

56  https://www.galipnuts.net/ (accessed 28 February 2021).

57  https://kakaduplumco.com/products/vital-c-with-kakadu-plum-and-desert-lime (accessed 28 February 2021).

58  Barker, G., editor. 2013. *Rainforest Foraging and Farming in Island Southeast Asia: The archaeology of the Niah Caves, Sarawak*, Vol. I. University of Cambridge.

59  Barker, G., H. Barton, M. Bird et al. 2007. The 'human revolution' in lowland tropical Southeast Asia: The antiquity and behavior of anatomically modern humans at Niah Cave (Sarawak, Borneo). *Journal of Human Evolution* **52**:243–61.

60  Denham, T., R. Fullagar and L. Head. 2009. Plant exploitation on Sahul.

61 Wedage, O., N. Amano, M.C. Langley et al. 2019. Specialized rainforest hunting by *Homo sapiens* ~45,000 years ago. *Nature Communications* **10**:739.

62 Summerhayes, G.R., J.H. Field, B. Shaw and D. Gaffney. 2017. The archaeology of forest exploitation and change in the tropics during the Pleistocene: The case of Northern Sahul (Pleistocene New Guinea). *Quaternary International* **448**:14–30.

63 Summerhayes, G.R., J.H. Field, B. Shaw and D. Gaffney. 2017. The archaeology of forest exploitation and change in the tropics during the Pleistocene.

64 Bird, R.B., N. Tayor, B.F. Codding and D.W. Bird. 2013. Niche construction and Dreaming logic: Aboriginal patch mosaic burning and varanid lizards (*Varanos gouldii*) in Australia. *Proceedings of the Royal Society B: Biological Sciences* **280**:20132297.

65 Docker, E.G. 1965. *Simply Human Beings*. Angus & Robertson.

66 Docker, E.G. 1965. *Simply Human Beings*.

67 Latest edition: Alexander, S. 2014. *The Cook's Companion: The complete book of ingredients and recipes*. Penguin.

68 Dixon, R.M.W. 2017. *Dyirbal Thesaurus and Dictionary across Ten Dialects: Part One nominals, nouns, adjectives and time words*. Language and Culture Research Centre, James Cook University.

69 Harris, D.R. 1987. Aboriginal subsistence in a tropical rain forest environment: Food procurement, cannibalism, and population regulation in northeastern Australia. pp. 357–86 in M. Harris and E.B. Ross, editors. *Food and Evolution: Toward a theory of human food habits*. Temple University Press.

70 Lumholtz, C. 1908. *Among Cannibals: An account of four years' travels in Australia and of camp life with the Aborigines of Queensland*. Charles Scribner's Sons.

71 Reeves, J.M., H.C. Bostock, L.K. Ayliffe et al. 2013. Palaeoenvironmental change in tropical Australasia over the last 30,000 years.

72 Hilbert, D.W., A. Graham and M.S. Hopkins. 2007. Glacial and interglacial refugia within a long-term rainforest refugium: The Wet Tropics Bioregion of NE Queensland, Australia. *Palaeogeography, Palaeoclimatology, Palaeoecology* **251**:104–18.

73 Ferrier, Å. 2015. *Journeys into the Rainforest: Archaeology of culture change and continuity on the Evelyn Tableland, North Queensland*. ANU Press.

74 Ferrier, Å. and R. Cosgrove. 2012. Aboriginal exploitation of toxic nuts as a late-Holocene subsistence strategy in Australia's tropical rainforests. pp. 103–20 in S.G. Haberle and B. David, editors. *Peopled Landscapes: Archaeological and biogeographic approaches to landscapes*. ANU Press.

75 Cited in Tuechler, A., Å. Ferrier and R. Cosgrove. 2014. Transforming the inedible to the edible: An analysis of the nutritional returns from Aboriginal nut processing in Queensland's Wet Tropics. *Australian Archaeology* **79**:26–33.

76 Cited in Tuechler, A., Å. Ferrier and R. Cosgrove. 2014. Transforming the inedible to the edible.

77 Steinberger, L. 2014. Hands in pockets: Cultural environments of the Atherton Tablelands of the past 1,500 years. PhD thesis. University of Queensland.

78 Hill, R. and A. Baird. 2003. Kuku-Yalanji rainforest Aboriginal people and carbohydrate resource management in the Wet Tropics of Queensland, Australia. *Human Ecology* **31**:27–52.

79 Crabtree, S.A., D.W. Bird and R.B. Bird. 2019. Subsistence transitions and the simplification of ecological networks in the Western Desert of Australia. *Human Ecology* **47**:165–77.

80 Roger Martin, personal communication, June 2021.

81 Savage, P. 1992. *Christie Palmerston, Explorer. Records of North Queensland History No 2*. James Cook University.

82    Rowe, C., M. Brand, L.B. Hutley et al. 2019. Holocene savanna dynamics in the seasonal tropics of northern Australia. *Review of Palaeobotany and Palynology* **267**: 17–31.

83    Rowe, C., M. Brand, L.B. Hutley et al. 2019. Holocene savanna dynamics in the seasonal tropics of northern Australia.

84    Rowe, C., M. Brand, L.B. Hutley et al. 2019. Holocene savanna dynamics in the seasonal tropics of northern Australia.

85    Fletcher, M.-S., T. Hall and A.N. Alexandra. 2021. The loss of an indigenous constructed landscape following British invasion of Australia: An insight into the deep human imprint on the Australian landscape. *Ambio* **50**:138–49.

86    Steinberger, L. 2014. Hands in pockets.

87    Stanton, P., M. Parsons, D. Stanton and M. Stott. 2014. Fire exclusion and the changing landscape of Queensland's Wet Tropics Bioregion 2: The dynamics of transition forests and implications for management. *Australian Forestry* **77**:58–68.

88    Stanton, P., M. Parsons, D. Stanton and M. Stott. 2014. Fire exclusion and the changing landscape of Queensland's Wet Tropics Bioregion 2.

89    Steinberger, L. 2014. Hands in pockets.

90    Stanton, P., D. Stanton, M. Stott and M. Parsons. 2014. Fire exclusion and the changing landscape of Queensland's Wet Tropics Bioregion 1: The extent and pattern of transition. *Australian Forestry* **77**:51–7.

### Chapter Nine: Thiaki's theatre of the absurd

1    Rose, D.B. 1996. *Nourishing Terrains: Australian Aboriginal views of landscape and wilderness*. Australian Heritage Commission.

2    Curr, E.M. 1880. *The Australian Race: Its origin, languages, customs, place of landing in Australia and the routes by which it spread itself over that continent*. Vol. 1. John Ferres, Government Printer.

3    http://palaeos.com/science/glossary.html#social_darwinism (accessed 16 May 2021).

4    Bottoms, T. 2013. *Conspiracy of Silence: Queensland's frontier killing times*. Allen & Unwin.

5    Cited in Birtles, T.G. 1967. A survey of land use, settlement and society in the Atherton-Evelyn district, north Queensland, 1880–1914. Master's thesis. University of Sydney.

6    Cook, J. 1893. *Captain Cook's Journal. First Voyage*. Captain W.J.L. Wharton, editor. Project Gutenberg Australia. http://gutenberg.net.au/ebooks/e00043.html.

7    Carron, W. 1849. *Narrative of an expedition undertaken under the direction of the late Mr Assistant Surveyor E.B. Kennedy for the exploration of the country lying between Rockingham Bay and Cape York*. Kemp and Fairfax.

8    A map of Leichhardt's journey can be found on http://adb.anu.edu.au/entity/8843.

9    Bottoms, T. 2015. *Cairns City of the South Pacific: A history 1770–1995*. Bunu Bunu Press.

10   Loos, N.A. 1980. Queensland's kidnapping Act: *The Native Labourers Protection Act of 1884*. *Aboriginal History* **4**:150–73.

11   Jones, D. 1961. *Cardwell Shire Story*. Jacaranda Press.

12   MacGillivray, J. 1852. *Narrative of the voyage of H.M.S. Rattlesnake: commanded by the late Captain Owen Stanley during the years 1846–1850, including discoveries and surveys in New Guinea, the Louisiade Archipelago, etc., to which is added the account of Mr. E.B. Kennedy's expedition for the exploration of the Cape York Peninsula*. T. & W. Boone. https://www.nma.gov.au/learn/encounters-education/community-stories/rockingham-bay (accessed 6 August 2022).

13  Jukes, J.B. 1847. *Narrative of the surveying voyage of H.M.S. Fly: commanded by Captain F.P. Blackwood, R.N., in Torres Strait, New Guinea, and other islands of the Eastern Archipelago, during the years 1842–1846*. Vol. 1. T. & W. Boone.

14  Carron, W. 1849. *Narrative of an expedition undertaken under the direction of the late Mr Assistant Surveyor E.B. Kennedy.*

15  https://apps.lucidcentral.org/rainforest/text/intro/index.html (accessed 1 February 2021).

16  Dixon, R.M.W. 2017. *Dyirbal Thesaurus and Dictionary across Ten Dialects: Part One nominals, nouns, adjectives and time words*. Language and Culture Research Centre, James Cook University.

17  https://noosasnativeplants.com.au/plant-details/?pId=447 (accessed 1 February 2021).

18  Rossetto, M., E.J. Ens, T. Honings et al. 2017. From songlines to genomes: Prehistoric assisted migration of a rain forest tree by Australian Aboriginal people. *PLoS One* **12**:e0186663.

19  Gammage, B. 2012. *The Biggest Estate on Earth: How Aborigines made Australia*. Allen & Unwin.

20  Stanner, W.E.H. 1965. Aboriginal territorial organization: Estate, range, domain and regime. *Oceania* **36**:1–26.

21  Gammage, B. 2012. *The Biggest Estate on Earth.*

22  The term was used in Gammage, B. 2012. *The Biggest Estate on Earth.*

23  Fitzpatrick, A., I.J. McNiven, J. Specht and S. Ulm. 2018. Stylistic analysis of stone arrangements supports regional cultural interactions along the northern Great Barrier Reef, Queensland. *Australian Archaeology* **84**:129–44.

24  Jack, R.L. 1921. *Northmost Australia: Three centuries of exploration, discovery, and adventure in and around the Cape York Peninsula, Queensland*. Simpkin, Marshall, Hamilton, Kent & Co.

25  Anderson, S. 2009. *Pelletier: The forgotten castaway of Cape York*. Melbourne Books.

26  Sutton, P. 1978. *Wik: Aboriginal society, territory and language at Cape Keerweer, Cape York Peninsula, Australia*. University of Queensland.

27  https://www.australiaonthemap.org.au/cape-york-and-torres-strait-1606-to-1643-2/ (accessed 1 February 2021).

28  Loos, N.A. 1974. Aboriginal–Dutch relations in North Queensland, 1606–1756. *Queensland Heritage* **3**:3–8.

29  Report to the Governors of the Dutch East India Company, *in* Jack, R.L. 1921. *Northmost Australia.* p. 61.

30  Loos, N.A. 1974. Aboriginal–Dutch relations in North Queensland, 1606–1756.

31  Jones, D. 1973. *Hurricane Lamps and Blue Umbrellas: The story of Innisfail and the Shire of Johnstone North Queensland*. G.K. Bolton Printers.

32  Cook, J. 1893. *Captain Cook's Journal.*

33  The Aboriginal side of the meeting comes from https://www.abc.net.au/radionational/programs/earshot/the-story-of-australias-first-reconciliation/6773096.

34  The Aboriginal side of the meeting comes from https://www.abc.net.au/radionational/programs/earshot/the-story-of-australias-first-reconciliation/6773096.

35  Jones, D. 1961. *Cardwell Shire Story.*

36  King, P.P. 1827. *Narrative of a survey of the intertropical and western coasts of Australia: performed between the years 1818 and 1822*. Entry for 21 June 1819. John Murray.

37  Jack, R.L. 1921. *Northmost Australia.*

38  Anderson, S. 2009. *Pelletier.*

39  Anderson, S. 2009. *Pelletier.*

40  Jones, D. 1961. *Cardwell Shire Story.*

41  http://www.bookrags.com/ebooks/16027/130.html#gsc.tab=0 (accessed 6 August 2021).

42  Stokes, J.L. 1846. *Discoveries in Australia; with an account of the coasts and rivers explored and surveyed during the voyage of the H.M.S. Beagle in the years 1837–1843.* Vol. 1. T. & W. Boone.

43  McNiven, I.J. 2019. Beyond bridge and barrier: Torres Strait and curious artefact distributions between Queensland and New Guinea. *Oceanic Arts Society Journal* **24**:3–6.

44  Jones, D. 1973. *Hurricane Lamps and Blue Umbrellas.*

45  Moore, D.R. 1979. *Islanders and Aborigines at Cape York: An ethnographic reconstruction based on the 1848–1850 'Rattlesnake' journals of O.W. Brierly and information he obtained from Barbara Thompson.* Australian Institute of Aboriginal Studies.

46  This type of arrangement (for other places along the east coast of Australia and in Tasmania) is described in Gammage, B. 2012. *The Biggest Estate on Earth.*

47  Moore, D.R. 1979. *Islanders and Aborigines at Cape York.*

48  Moore, D.R. 1979. *Islanders and Aborigines at Cape York.*

49  Sharp, N. 1992. *Footprints Along the Cape York Sandbeaches.* Aboriginal Studies Press.

50  Cited in Sharp, N. 1992. *Footprints Along the Cape York Sandbeaches.*

## Chapter Ten: Evolution, interrupted

1  Gammage, B. 2012. *The Biggest Estate on Earth: How Aborigines made Australia.* Allen & Unwin. p. 95.

2  Evans, R. and B. Thorpe. 2001. Indigenocide and the massacre of Aboriginal history. *Overland* **163**:21–39.

3  Reynolds, H. 1974. Racial thought in early colonial Australia. *Australian Journal of Politics & History* **20**:45–53.

4  Loos, N. 1976. Aboriginal–European relations in north Queensland, 1861–1897. PhD thesis. James Cook University.

5  https://australianfrontierconflicts.com.au/timelines/timeline-of-australian-frontier-conflicts/ (accessed 27 May 2021).

6  Cook, J. 1893. *Captain Cook's Journal. First Voyage.* Captain W.J.L. Wharton, editor. Project Gutenberg Australia. http://gutenberg.net.au/ebooks/e00043.html.

7  Burke, H., B. Barker, L. Wallis et al. 2020. Betwixt and between: Trauma, survival and the Aboriginal troopers of the Queensland Native Mounted Police. *Journal of Genocide Research* **22**:317–33.

8  Evans, R. and R. Ørsted-Jensen. 2014. 'I cannot say the numbers that were killed': Assessing violent mortality on the Queensland frontier. *Social Science Research Network* **2467836**:1–11.

9  P.D.T. (17 June 1871) cited in Loos, N. 1976. Aboriginal–European relations in north Queensland, 1861–1897.

10  C.C. (1877) cited in Loos, N. 1976. Aboriginal–European relations in north Queensland, 1861–1897.

11  Jones, D. 1973. *Hurricane Lamps and Blue Umbrellas: The story of Innisfail and the Shire of Johnstone North Queensland.* G.K. Bolton Printers.

12  Dalrymple, G.E. 1874. *Narrative and Reports of the Queensland North-East Coast Expedition, 1873.* Queensland Government Printer, Brisbane.

13   Hynes, R.A. and A.K. Chase. 1982. Plants, sites and domiculture: Aboriginal influence upon plant communities in Cape York Peninsula. *Archaeology in Oceania* **17**:38–50.
14   Evans, R. and B. Thorpe. 2001. Indigenocide and the massacre of Aboriginal history.
15   Genever, G. 2006. *Failure of Justice: The story of the Irvinebank massacre*. Eacham Historical Society.
16   This was reported by Johnstone as an appendix in Dalrymple, G.E. 1874. *Narrative and Reports of the Queensland North-East Coast Expedition, 1873*. p. 14.
17   Evans, R. and B. Thorpe. 2001. Indigenocide and the massacre of Aboriginal history.
18   Jones, D. 1973. *Hurricane Lamps and Blue Umbrellas*.
19   Jones, D. 1973. *Hurricane Lamps and Blue Umbrellas*.
20   Evans, R. and B. Thorpe. 2001. Indigenocide and the massacre of Aboriginal history.
21   Dalrymple, G.E. 1874. *Narrative and Reports of the Queensland North-East Coast Expedition, 1873*. p. 14.
22   Dalrymple, G.E. 1874. *Narrative and Reports of the Queensland North-East Coast Expedition, 1873*. p. 16.
23   Bottoms, T. 2015. *Cairns City of the South Pacific: A history 1770–1995*. Bunu Bunu Press.
24   Dalrymple, G.E. 1874. *Narrative and Reports of the Queensland North-East Coast Expedition, 1873*. pp. 14, 18.
25   This section combines information from Dalrymple, G.E. 1874. *Narrative and Reports of the Queensland North-East Coast Expedition, 1873*, and from DES. 2019. A Biodiversity Planning Assessment for the Wet Tropics Bioregion: Expert Panel Report. Version 1.1. Department of Environment and Science, Queensland Government.
26   Dalrymple, G.E. 1874. *Narrative and Reports of the Queensland North-East Coast Expedition, 1873*. p. 21.

## Chapter Eleven: The other side of the pincer

1   Roberts, S.H. 1968. *History of Australian Land Settlement 1788–1920*. Macmillan.
2   Roberts, S.H. 1968. *History of Australian Land Settlement 1788–1920*.
3   Cited in Loos, N. 1976. Aboriginal–European relations in north Queensland, 1861–1897. PhD thesis. James Cook University.
4   Loos, N. 1976. Aboriginal–European relations in north Queensland, 1861–1897.
5   Loos, N. 1976. Aboriginal–European relations in north Queensland, 1861–1897.
6   Burke, H., B. Barker, N. Cole et al. 2018. The Queensland Native Police and strategies of recruitment on the Queensland frontier, 1849–1901. *Journal of Australian Studies* **42**:297–313.
7   Loos, N. 1976. Aboriginal–European relations in north Queensland, 1861–1897.
8   Burke, H., B. Barker, N. Cole et al. 2018. The Queensland Native Police and strategies of recruitment on the Queensland frontier, 1849–1901.
9   Burke, H., B. Barker, L. Wallis et al. 2020. Betwixt and between: Trauma, survival and the Aboriginal troopers of the Queensland Native Mounted Police. *Journal of Genocide Research* **22**:317–33.
10   Burke, H., B. Barker, N. Cole et al. 2018. The Queensland Native Police and strategies of recruitment on the Queensland frontier, 1849–1901, p. 308.
11   Docker, E.G. 1965. *Simply Human Beings*. Angus & Robertson.
12   Barker, B., L.A. Wallis, H. Burke et al. 2020. The archaeology of the 'Secret War': The material evidence of conflict on the Queensland frontier, 1849–1901. *Queensland Archaeological Research* **23**:25–41.

13   Barker, B., L.A. Wallis, H. Burke et al. 2020. The archaeology of the 'Secret War'.
14   Leichhardt, L. 2002. *Journal of an overland expedition in Australia: From Moreton Bay to Port Essington, a distance of upwards of 3000 miles, during the years 1844–1845.* University of Sydney Library (prepared from the print edition published by T. & W. Boone. 1847).
15   Loos, N. 1976. Aboriginal–European relations in north Queensland, 1861–1897.
16   E. Palmer cited in Loos, N. 1976. Aboriginal–European relations in north Queensland, 1861–1897.
17   *Port Denison Times* (4 June 1868) cited in Evans, R. and B. Thorpe. 2001. Indigenocide and the massacre of Aboriginal history. *Overland* **163**:21–40.
18   Jones, D. 1961. *Cardwell Shire Story.* The Jacaranda Press.
19   This version is from Loos, N. 1976. Aboriginal–European relations in north Queensland, 1861–1897. Loos had sighted Morrill's diary of the event.
20   Dalrymple, J.E. 1865. On the new settlement in Rockingham Bay, and advance of colonization over north-eastern Australia; including Mr. J.E. Dalrymple's report on his journey from Rockingham Bay to the Valley of Lagoons. *The Journal of the Royal Geographical Society of London* **35**:191–212.
21   Dalrymple, J.E. 1865. On the new settlement in Rockingham Bay, and advance of colonization over north-eastern Australia.
22   Morrill, J. 2006. *17 Years Wandering Among the Aboriginals.* David M. Welch. p. 14.
23   https://en.wikipedia.org/wiki/Biria_people (accessed 4 May 2021).
24   https://press-files.anu.edu.au/downloads/press/n4525/html/ch02.xhtml#footnote-194 (accessed 4 May 2021).
25   Morrill, J. 2006. *17 Years Wandering Among the Aboriginals.* p. 14.
26   Cited in https://press-files.anu.edu.au/downloads/press/n4525/html/ch02.xhtml#footnote-188-backlink (accessed 5 May 2021).

## Chapter Twelve: Desultory little massacres

1   *Cooktown Courier* (1878), referring to the 'lack of decent massacres' at the mining frontier of the Palmerston and Hodgkinson, cited in Loos, N. 1976. Aboriginal–European relations in north Queensland, 1861–1897. PhD thesis. James Cook University. p. 215.
2   Jones, D. 1973. *Hurricane Lamps and Blue Umbrellas: The story of Innisfail and the Shire of Johnstone North Queensland.* G.K. Bolton Printers.
3   Loos, N. 1976. Aboriginal–European relations in north Queensland, 1861–1897.
4   Taylor, P. 1994. *Growing Up: Forestry in Queensland.* Allen & Unwin.
5   Jones, D. 1973. *Hurricane Lamps and Blue Umbrellas.*
6   Jones, D. 1973. *Hurricane Lamps and Blue Umbrellas.*
7   Jones, D. 1973. *Hurricane Lamps and Blue Umbrellas.* p. 117.
8   Lumholtz, C. 1908. *Among Cannibals: An account of four years' travels in Australia and of camp life with the Aborigines of Queensland.* Charles Scribner's Sons. p. 89.
9   Lumholtz, C. 1908. *Among Cannibals.* p. 263.
10  Lumholtz, C. 1908. *Among Cannibals.* p. 158.
11  Lumholtz, C. 1908. *Among Cannibals.* p. 189.
12  Jones, D. 1961. *Cardwell Shire Story.* The Jacaranda Press.
13  Preece, N.D. 2013. Tangible evidence of historic Australian indigenous savanna management. *Austral Ecology* **38**:241–50.
14  Steinberger, L. 2014. Hands in pockets: Cultural environments of the Atherton Tablelands of the past 1,500 years. PhD thesis. University of Queensland. p. 59.

15  Pike, G. 1951. James Venture Mulligan: Prospector and explorer of the north. *Journal of the Royal Historical Society of Queensland* 4:494–508.

16  Birtles, T.G. 1967. A survey of land use, settlement and society in the Atherton-Evelyn district, north Queensland, 1880–1914. Master's thesis. University of Sydney.

17  Loos, N. 1976. Aboriginal–European relations in north Queensland, 1861–1897.

18  Cited in *Expedition in Search of Gold and other Minerals in the Palmer districts, by Mulligan and Party 29th April – 23rd September 1875.* Eacham Historical Society.

19  Pike, G. 1951. James Venture Mulligan.

20  Savage, P. 1992. *Christie Palmerston, Explorer. Records of North Queensland History No 2.* James Cook University.

21  Dixon, R.M.W. 1972. Dyirbal: The language and its speakers. pp. 22–38 in R.M.W. Dixon, editor. *The Dyirbal Language of North Queensland.* Cambridge University Press.

22  Birtles, T. 1982. Trees to burn: Settlement in the Atherton-Evelyn rainforest, 1880–1900. *North Australian Research Bulletin* 8:31–51.

23  Birtles, T.G. 1967. A survey of land use, settlement and society in the Atherton-Evelyn district, north Queensland, 1880–1914.

24  Birtles, T.G. 1967. A survey of land use, settlement and society in the Atherton-Evelyn district, north Queensland, 1880–1914. p. 55.

25  Birtles, T. 1982. Trees to burn.

26  Steinberger, L. 2014. Hands in pockets.

27  Savage, P. 1992. *Christie Palmerston, Explorer.*

28  Meston, A. 1896. Report on the Aboriginals of Queensland. https://nla.gov.au/nla.obj-52864172 (accessed 10 February 2019).

29  Savage, P. 1992. *Christie Palmerston, Explorer.*

30  Douglas, A.D. 1882. Herberton to Mourilyan (part 2). *The Queenslander* (1 July 1882). pp. 12–13.

31  Birtles, T.G. 1967. A survey of land use, settlement and society in the Atherton-Evelyn district, north Queensland, 1880–1914.

32  Savage, P. 1992. *Christie Palmerston, Explorer.*

33  Birtles, T.G. 1967. A survey of land use, settlement and society in the Atherton-Evelyn district, north Queensland, 1880–1914.

34  Frawley, K.J. 1983. Forest and land management in north-east Queensland: 1859–1960. PhD thesis. ANU.

35  Loos, N. 1976. Aboriginal–European relations in north Queensland, 1861–1897.

36  Birtles, T.G. 1967. A survey of land use, settlement and society in the Atherton-Evelyn district, north Queensland, 1880–1914.

37  Birtles, T.G. 1967. A survey of land use, settlement and society in the Atherton-Evelyn district, north Queensland, 1880–1914.

38  Birtles, T. 1978. Changing perception and response to the Atherton-Evelyn rainforest environment 1880–1920. *Proceedings of the 15th Annual Conference.* Institute of Australian Geographers.

39  Eacham Historical Society. 1995. *Malanda: In the Shadow of Bartle Frere.* Eacham Historical Society.

40  Birtles, T. 1967. A survey of land use, settlement and society in the Atherton-Evelyn district, north Queensland, 1880–1914.

41  Cited in Birtles, T. 1978. Changing perception and response to the Atherton-Evelyn rainforest environment 1880–1920.

42  Cited in Gilmore, M. 2005. Kill, cure or strangle: The history of government intervention in three key agricultural industries on the Atherton Tablelands, 1895–2005. PhD thesis.

James Cook University. The estimate of people is from Walter Roth, the first northern protector of Aboriginal people.

43    Gilmore, M. 2005. Kill, cure or strangle.

44    Genever, G. 2006. *Failure of Justice: The story of the Irvinebank massacre.* Eacham Historical Society.

45    Loos, N. 1976. Aboriginal–European relations in north Queensland, 1861–1897. p. 191.

46    Genever, G. 2006. *Failure of Justice.*

47    Loos, N. 1976. Aboriginal–European relations in north Queensland, 1861–1897.

48    Cited in Loos, N. 1976. Aboriginal–European relations in north Queensland, 1861–1897. p. 285.

49    Loos, N. 1976. Aboriginal–European relations in north Queensland, 1861–1897.

50    Cited in Loos, N. 1976. Aboriginal–European relations in north Queensland, 1861–1897. p. 673.

51    Meston, A. 1896. Report on the Aboriginals of Queensland. p. 8.

52    Loos, N. 1982. *Invasion and Resistance: Aboriginal–European relations on the North Queensland frontier 1861–1897.* ANU Press. p. 176.

53    Parry-Okeden, W.E. 1897. North Queensland Aborigines and Native Police. *The Brisbane Courier* (16 April 1897). p. 6. https://trove.nla.gov.au/newspaper/article/3649161.

54    *Aboriginals Protection and Restriction of the Sale of Opium Act 1897,* s. 3, Interpretation.

55    Cited in Loos, N. 1976. Aboriginal–European relations in north Queensland, 1861–1897. p. 694.

56    Burke, H., B. Barker, L. Wallis et al. 2020. Betwixt and between: Trauma, survival and the Aboriginal troopers of the Queensland Native Mounted Police. *Journal of Genocide Research* **22**:317–33.

57    Crome, F.H.J., L.A. Moore and G.C. Richards. 1992. A study of logging damage in upland rainforest in north Queensland. *Forest Ecology and Management* **49**:1–29.

## Chapter Thirteen: Fugitive pieces

1    Webb, L. 1963, 'Trees are your friends', *Wildlife Australia* **1**·10.

2    Birtles, T. 1982. Trees to burn: Settlement in the Atherton-Evelyn rainforest, 1880–1900. *North Australian Research Bulletin* **8**:31–51.

3    *Barron Valley Advocate and Mareeba Advertiser* (31 May 1905) cited in Birtles, T. 1978. Changing perception and response to the Atherton-Evelyn rainforest environment 1880–1920. *Proceedings of the 15th Annual Conference.* Institute of Australian Geographers.

4    *Barron Valley Advocate and Mareeba Advertiser* (31 May 1905) cited in Birtles, T. 1978. Changing perception and response to the Atherton-Evelyn rainforest environment 1880–1920.

5    Statham, A. 1998. *Cows in the Vine Scrub: A history of dairying on the Atherton Tableland.* Malanda Dairyfoods.

6    Statham, A. 1998. *Cows in the Vine Scrub.*

7    Cited in Statham, A. 1998. *Cows in the Vine Scrub.* p. 27.

8    Frawley, K.J. 1983. Forest and land management in north-east Queensland: 1859–1960. PhD thesis. ANU.

9    Frawley, K.J. 1983. Forest and land management in north-east Queensland: 1859–1960.

10   https://adb.anu.edu.au/biography/hill-walter-12981 (accessed 7 June 2021).

11  Frawley, K.J. 1983. Forest and land management in north-east Queensland: 1859–1960.
12  https://adb.anu.edu.au/biography/mcilwraith-sir-thomas-4099 (accessed 8 September 2021).
13  Vanclay, J.K. 1996. Lessons from the Queensland rainforests: Steps towards sustainability. *Journal of Sustainable Forestry* **3**:1–27.
14  Cited in Taylor, P. 1994. *Growing Up: Forestry in Queensland.* Allen & Unwin. p. 50.
15  Taylor, P. 1994. *Growing Up.*
16  Swain (1931) cited in Frawley, K.J. 1983. Forest and land management in north-east Queensland: 1859–1960. p. 152.
17  Statham, A. 1998. *Cows in the Vine Scrub.*
18  Frawley, K.J. 1983. Forest and land management in north-east Queensland: 1859–1960.
19  Calculated from the land being £3 per acre in 1911 using https://www.in2013dollars. com/uk/inflation/1911?amount=1 as my starting point.
20  Birtles, T.G. 1967. A survey of land use, settlement and society in the Atherton-Evelyn district, north Queensland, 1880–1914. Master's thesis. University of Sydney.
21  Cited in Statham, A. 1998. *Cows in the Vine Scrub.*
22  Eacham Historical Society notes on Billy Barlow and the Thomas family.
23  Frawley, K.J. 1983. Forest and land management in north-east Queensland: 1859–1960.
24  Frawley, K.J. 1983. Forest and land management in north-east Queensland: 1859–1960.
25  Cilento (1925) cited in Statham, A. 1998. *Cows in the Vine Scrub.* p. 187.
26  Grenning (1979) cited in Frawley, K.J. 1983. Forest and land management in north-east Queensland: 1859–1960.
27  Statham, A. 1998. *Cows in the Vine Scrub.*
28  Gilmore, M. 2005. Kill, cure or strangle: The history of government intervention in three key agricultural industries on the Atherton Tablelands, 1895–2005. PhD thesis. James Cook University.
29  Swain (1930) cited in Frawley, K.J. 1983. Forest and land management in north-east Queensland: 1859–1960. p. 223.
30  Frawley, K.J. 1983. Forest and land management in north-east Queensland: 1859–1960.
31  The Royal Commission on the Development of North Queensland (Land Settlement and Forestry). 1931. *Report.*
32  Frawley, K.J. 1983. Forest and land management in north-east Queensland: 1859–1960. p. 227.
33  The Royal Commission on the Development of North Queensland (Land Settlement and Forestry). 1931. *Report.* Appendix 6.
34  The Royal Commission on the Development of North Queensland (Land Settlement and Forestry). 1931. *Report.* p. 34.
35  Cited in Statham, A. 1998. *Cows in the Vine Scrub.* p. 176.
36  The Royal Commission on the Development of North Queensland (Land Settlement and Forestry). 1931. *Report.* p. 17.
37  The Royal Commission on the Development of North Queensland (Land Settlement and Forestry). 1931. *Report.* p. 8.
38  Taylor, P. 1994. *Growing Up.*
39  Frawley, K.J. 1983. Forest and land management in north-east Queensland: 1859–1960. p. 200.
40  Taylor, P. 1994. *Growing Up.*
41  Frawley, K.J. 1983. Forest and land management in north-east Queensland: 1859–1960. p. 200.
42  Statham, A. 1998. *Cows in the Vine Scrub.* p. 253.

43  Cited in Frawley, K.J. 1987. *The Maalan Group Settlement, North Queensland, 1954: An historical geography.* Monograph Series No. 2. Department of Geography and Oceanography UNSW, Australian Defence Academy.

44  Frawley, K.J. 1983. Forest and land management in north-east Queensland: 1859–1960.

45  Cited in Frawley, K.J. 1987. *The Maalan Group Settlement, North Queensland, 1954.* p. 51.

46  A settler cited in Frawley, K.J. 1987. *The Maalan Group Settlement, North Queensland, 1954.* p. 80.

47  Gilmour and Reilly (1970) cited in Frawley, K.J. 1987. *The Maalan Group Settlement, North Queensland, 1954.*

48  https://www.queenslandcountrylife.com.au/story/7034024/tablelands-dairy-industry-optimistic/?cs=4733 (accessed 1 July 2022).

49  Dawson, C. 1969. American family owns chunk of Australia. *The Sunday Mail Color Magazine.* 12 October.

50  https://www.texasmonthly.com/travel/the-last-empire/ (accessed 2 September 2021).

51  Dixon, R.M.W. 2017. *Dyirbal Texts: 78 legends, stories, autobiographies, conversations, and remedies in Jirrbal, Girramay, Mamu, and Gulngay.* Language and Culture Research Centre, James Cook University. I have taken pieces of the conversation.

## Chapter Fourteen: The final stand?

1   Wilkie, B. 2017. *The Daintree Blockade: The battle for Australia's tropical rainforests.* Four Mile Books.

2   Collins, M.J.B. 1994. Patterns and rates of rainforest conversion on the Atherton and Evelyn Tablelands, Northeastern Queensland, 1978–1988. *Proceedings of the Royal Society of Queensland* **104**:1–10.

3   Cited in Frawley, K.J. 1983. Forest and land management in north-east Queensland: 1859–1960. PhD thesis. ANU.

4   Wilkie, B. 2017. *The Daintree Blockade.*

5   DES. 2019. A Biodiversity Planning Assessment for the Wet Tropics Bioregion: Expert Panel Report. Version 1.1. Department of Environment and Science, Queensland Government.

6   https://www.abc.net.au/news/2019-08-17/terania-creek-anti-logging-protest-40-years-on/11406660?nw=0&r=Gallery (accessed 3 September 2021).

7   Crome, F.H.J., L.A. Moore and G.C. Richards. 1992. A study of logging damage in upland rainforest in north Queensland. *Forest Ecology and Management* **49**:1–29.

8   DES. 2019. A Biodiversity Planning Assessment for the Wet Tropics Bioregion.

9   Valentine, P.S. and R. Hill. 2008. The establishment of a World Heritage Area. pp. 81–93 in N.E. Stork and S.M. Turton, editors. *Living in a Dynamic Tropical Forest Landscape.* Blackwell Publishing.

10  Taylor, P. 1994. *Growing Up: Forestry in Queensland.* Allen & Unwin.

11  Taylor, P. 1994. *Growing Up.* p. 223.

12  Frawley, K.J. 1983. Forest and land management in north-east Queensland: 1859–1960. p. 322.

13  https://www.dcceew.gov.au/sites/default/files/documents/190505.pdf (accessed 7 August 2022).

14  Driml, S. 1994. *Protection for Profit: Economic and financial values of the Great Barrier Reef World Heritage Area and other protected areas.* Great Barrier Reef Marine Park Authority.

15  Valentine, P.S. and R. Hill. 2008. The establishment of a World Heritage Area.

16  Rainforest Conservation Society of Queensland. 1986. *Tropical Rainforests of North Queensland: Their conservation significance.* AGPS.

17  History of the Wet Tropics World Heritage Area. https://cafnec.org.au/about/how-the-wet-tropics-was-won/.

18  Le Saout, S., M. Hoffmann, Y. Shi et al. 2013. Protected areas and effective biodiversity conservation. *Science* **342**:803–5.

19  Stanton, P., D. Stanton, M. Stott and M. Parsons. 2014. Fire exclusion and the changing landscape of Queensland's Wet Tropics Bioregion 1. The extent and pattern of transition. *Australian Forestry* **77**:51–7.

## Chapter Fifteen: From little things . . .

1  Rose, D.B. 1988. Exploring an Aboriginal land ethic. *Meanjin* **47**:378–87.

2  Pannell, S. 2006. *Yamani Country: A spatial history of the Atherton Tableland, North Queensland.* Rainforest CRC.

3  Cultural Values Project Steering Committee. 2016. 'Which way Australia's rainforest culture': Relisting the cultural values for world heritage. Discussion paper about realising the national and international recognition of the Rainforest Aboriginal cultural values of the Wet Tropics region and World Heritage Area. Compiled by Ro Hill, Ellie Bock and Petina Pert with and on behalf of the Rainforest Aboriginal Peoples and the Cultural Values Project Steering Committee.

4  https://news.mongabay.com/2019/08/indigenous-managed-lands-found-to-harbor-more-biodiversity-than-protected-areas/ (accessed 22 September 2021).

5  Kooyman, R.M., M. Rossetto, H. Sauquet and S.W. Laffan. 2013. Landscape patterns in rainforest phylogenetic signal: Isolated islands of refugia or structured continental distributions? *PLoS One* **8**:e80685.

6  Kooyman, R.M., P. Wilf, V.D. Barreda et al. 2014. Paleo-Antarctic rainforest into the modern Old World tropics: The rich past and threatened future of the 'southern wet forest survivors'. *American Journal of Botany* **101**:2121–35.

7  https://www.washingtonpost.com/science/2019/09/26/connecting-fractured-habitats-has-long-lasting-ecological-benefits-study-finds/?outputType=amp (accessed 19 October 2020).

8  Williams, S.E. and A. de la Fuente. 2021. Long-term changes in populations of rainforest birds in the Australia Wet Tropics bioregion: A climate-driven biodiversity emergency. *PLoS One* **16**:e0254307.

9  Williams, S.E., E.E. Bolitho and S. Fox. 2003. Climate change in Australian tropical rainforests: An impending environmental catastrophe. *Proceedings of the Royal Society B* **270**:1887–92.

10  Wintle, B.A., H. Kujala, A. Whitehead et al. 2019. Global synthesis of conservation studies reveals the importance of small habitat patches for biodiversity. *Proceedings of the National Academy of Sciences* **116**:909–14.

11  Damschen, E.I., L.A. Brudvig, M.A. Burt et al. 2019. Ongoing accumulation of plant diversity through habitat connectivity in an 18-year experiment. *Science* **365**:1478–80.

12  Stewart, F.E.C., S. Darlington, J.P. Volpe et al. 2019. Corridors best facilitate functional connectivity across a protected area network. *Scientific Reports* **9**:10852.

13  Laurance, W.F., S.G. Laurance and D.W. Hilbert. 2008. Long-term dynamics of a fragmented rainforest mammal assemblage. *Conservation Biology* **22**:1154–64.

14 Jones, D. 2016. Avifaunal community composition in a tropical forest corridor: A case study from the Atherton Tableland, North Queensland. *Independent Study Project (ISP) Collection*. 2303.

15 Campbell, A., J. Alexandra and D. Curtis. 2017. Reflections on four decades of land restoration in Australia. *The Rangeland Journal* **39**:405–16.

16 Tucker, N. 2018. A brief history of the Peterson Creek Corridor. *in Eacham Historical Society 2021 Bulletin No. 476*.

17 Tucker, N.I.J. and T. Simmons. 2009. Restoring a rainforest habitat linkage in north Queensland: Donaghy's Corridor. *Ecological Management & Restoration* **10**: 98–112.

18 Damschen, E.I., L.A. Brudvig, M.A. Burt et al. 2019. Ongoing accumulation of plant diversity through habitat connectivity in an 18-year experiment.

19 https://theconversation.com/global-climate-game-abandons-biodiversity-16570 (accessed 23 September 2021).

20 The United Nations Framework Convention on Climate Change (UNFCCC) and the Convention on Biological Diversity (CBD).

21 https://theconversation.com/australia-is-a-global-top-ten-deforester-and-queensland-is-leading-the-way-87259 (accessed 23 September 2023).

22 van Oosterzee, P. 2013. Global climate change abandons biodiversity. https://theconversation.com/global-climate-game-abandons-biodiversity-16570 (accessed 23 September 2021).

23 Exact figures are not known. A good summary can be found on mongabay (https://rainforests.mongabay.com/deforestation/), and for sequestration (https://www.qld.gov.au/environment/plants-animals/habitats/regrowth/regrowth-guides/rainforest/rainforest-carbon), and for Australia's emissions (https://www.industry.gov.au/news/australias-greenhouse-gas-emissions-march-2021-quarterly-update).

24 Waldron, A., D.C. Miller, D. Redding et al. 2017. Reductions in global biodiversity loss predicted from conservation spending. *Nature* 551.364–7.

25 Samuel, G. 2020. Independent review of the EPBC Act – Final report. Department of Agriculture, Water and the Environment.

26 Samuel, G. 2020. Independent review of the EPBC Act – Final report.

27 Australian Conservation Foundation. 2021. Federal government spending on Australia's environment and climate. https://d3n8a8pro7vhmx.cloudfront.net/auscon/pages/18803/attachments/original/1620346645/Federal_Environment_and_Climate_Budget_Analysis_-_May_2021.pdf?1620346645.

28 Bradford, M. and H.T. Murphy. 2019. The importance of large-diameter trees in the wet tropical rainforests of Australia. *PLoS One* **14**:e0208377.

29 ForestPlots.net, C. Blundo, J. Carilla et al. 2021. Taking the pulse of Earth's tropical forests using networks of highly distributed plots. *Biological Conservation* **260**:108849.

30 https://www.acf.org.au/budget_2018_19_investment_in_a_healthy_environment_cut_to_bare_bones_while_fossil_fuel_subsidies_continue (accessed 22 September 2021).

31 Tucker, N.I.J. and T. Simmons. 2009. Restoring a rainforest habitat linkage in north Queensland: Donaghy's Corridor.

32 DES. 2019. A Biodiversity Planning Assessment for the Wet Tropics Bioregion: Expert Panel Report. Version 1.1. Department of Environment and Science, Queensland Government.

33 Campbell, A., J. Alexandra and D. Curtis. 2017. Reflections on four decades of land restoration in Australia.

34 McGrath, C. 2007. End of broadscale clearing in Queensland. *Environmental and Planning Law Journal* **24**:5–13.

35 van Oosterzee, P., A. Dale and N.D. Preece. 2014. Integrating agriculture and climate change mitigation at landscape scale: Implications from an Australian case study. *Global Environmental Change* **29**:306–17.

36 http://www.millenniumassessment.org/documents/document.356.aspx.pdf (accessed 19 October 2021).

37 https://registry.verra.org/app/projectDetail/CCB/1581 (accessed 19 October 2021).

38 A riveting account is found in Wilkinson, M. 2020. *The Carbon Club: How a network of influential climate sceptics, politicians and business leaders fought to control Australia's climate policy.* Allen & Unwin.

39 Wilkinson, M. 2020. *The Carbon Club.*

40 Nelson, T., J. Nelson, J. Ariyaratnam and S. Camroux. 2013. An analysis of Australia's large scale renewable energy target: Restoring market confidence. *Energy Policy* **62**:386–400.

41 Miyamoto, M. and K. Takeuchi. 2019. Climate agreement and technology diffusion: Impact of the Kyoto Protocol on international patent applications for renewable energy technologies. *Energy Policy* **129**:1331–8.

42 European Parliament. 2019. European policies on climate and energy towards 2020, 2030 and 2050. https://www.europarl.europa.eu/RegData/etudes/BRIE/2019/631047/IPOL_BRI(2019)631047_EN.pdf.

43 World Economic Forum. 2020. Nature risk rising: Why the crisis engulfing nature matters for business and the economy. https://www.weforum.org/reports/nature-risk-rising-why-the-crisis-engulfing-nature-matters-for-business-and-the-economy.

44 Dasgupta, P. 2021. *The Economics of Biodiversity: The Dasgupta Review.* HM Treasury.

45 Costanza, R., R. d'Arge, R. de Groot et al. 1997. The total value of the world's ecosystem services and natural capital. *Nature* **387**:253–60.

46 Dasgupta, P. 2021. *The Economics of Biodiversity.*

47 Press release. https://www.ipcc.ch/2021/08/09/ar6-wg1-20210809-pr/ (accessed 24 October 2021).

48 Pörtner, H.-O., R.J. Scholes, J. Agard et al. 2021. IPBES-IPCC co-sponsored workshop report on biodiversity and climate change. IPBES and IPCC.

49 https://ipbes.net/news/Media-Release-Global-Assessment (accessed 21 October 2021).

50 Grantham, H.S., A. Duncan, T.D. Evans et al. 2020. Anthropogenic modification of forests means only 40% of remaining forests have high ecosystem integrity. *Nature Communications* **11**:5978.

51 UNFCCC. 2015. The Paris Agreement. United Nations FCCC/CP/2015/L.9.

52 ClimateWorks Australia. 2020. *Decarbonisation futures: Solutions, actions and benchmarks for net zero emissions Australia.*

53 Cited in United Nations Department of Economic and Social Affairs. 2021. *The Global Forest Goals Report.* United Nations Forum on Forests Secretariat.

54 Griscom, B.W., J. Adams, P.W. Ellis et al. 2017. Natural climate solutions. *Proceedings of the National Academy of Sciences* **114**:11645–50.

55 http://desdemonadespair.net/2018/10/scientists-say-halting-deforestation.html (accessed 25 October 2021).

56 https://unfccc.int/sites/default/files/resource/cma2021_L16_adv.pdf (accessed 29 October 2021).

57 https://unfccc.int/sites/default/files/resource/cma2021_L16_adv.pdf. Section IV, point 38. (accessed 29 October 2021).

58 https://www.weforum.org/projects/nature-action-agenda (accessed 29 October 2021).

59 Vivid Economics. 2021. *The Urgency of Biodiversity Action.* HM Treasury.

60 Swiss Re Institute. 2020. *Biodiversity and Ecosystem Services: A business case for re/insurance.*

61 Seddon, N., E. Daniels, R. Davis et al. 2020. Global recognition of the importance of nature-based solutions to the impacts of climate change. *Global Sustainability* **3**:np.

62 United Nations Environment Programme. 2021. *Making Peace with Nature: A scientific blueprint to tackle the climate, biodiversity and pollution emergencies.* Nairobi.

63 Sangha, K.K., J. Evans, A. Edwards et al. 2021. Assessing the value of ecosystem services delivered by prescribed fire management in Australian tropical savannas. *Ecosystem Services* **51**:101343.

64 https://www.qld.gov.au/environment/climate/climate-change/land-restoration-fund/about/overview (accessed 29 October 2021).

65 https://eol.org/pages/2906822/articles (accessed 29 October 2021).

66 Kooyman, R.M., R.J. Morley, D.M. Crayn et al. 2019. Origins and assembly of Malesian rainforests. *Annual Review of Ecology, Evolution, and Systematics* **50**:119–43.

67 Ritmejerytė, E., B.A. Boughton, M.J. Bayly and R.E. Miller. 2020. Unique and highly specific cyanogenic glycoside localization in stigmatic cells and pollen in the genus *Lomatia* (Proteaceae). *Annals of Botany* **126**:387–400.

68 Wet Tropics Management Authority. 2021. *State of Wet Tropics Report 2020–2021: Growing opportunities—landscape restoration for biodiversity and ecosystem recovery.* Cairns, Australia.

69 van Oosterzee, P., H. Liu and N.D. Preece. 2020. Cost benefits of forest restoration in a tropical grazing landscape: Thiaki rainforest restoration project. *Global Environmental Change* **63**:102105.

70 Pörtner, H.-O., R.J. Scholes, J. Agard et al. 2021. IPBES–IPCC co-sponsored workshop report on biodiversity and climate change.

71 Kooyman, R.M., P. Wilf, V.D. Barreda et al. 2014. Paleo-Antarctic rainforest into the modern Old World tropics.

72 Kooyman, R.M., M. Rossetto, H. Sauquet and S.W. Laffan. 2013. Landscape patterns in rainforest phylogenetic signal.

73 Lovejoy, T.E. and L. Hannah. 2018. Avoiding the climate fallsafe point. *Science Advances* **4**:eaau9981.

# INDEX